教育部－浪潮集团产学合作协同育人项目成果

普通高等学校计算机教育“十三五”规划教材

inspur 浪潮

SSM
框架应用开发与案例实战
（Spring+Spring MVC+MyBatis）

慕课版

浪潮优派◎策划

姬忠红 曹慧 周业勤◎主编

代敏 李海斌 刘丛丛 张新健 林培光◎副主编

人民邮电出版社

北京

图书在版编目（CIP）数据

SSM框架应用开发与案例实战 ：Spring+Spring MVC+MyBatis ：慕课版 / 姬忠红，曹慧，周业勤主编. -- 北京 ：人民邮电出版社，2021.5（2023.1重印）
普通高等学校计算机教育“十三五”规划教材
ISBN 978-7-115-53486-6

Ⅰ. ①S… Ⅱ. ①姬… ②曹… ③周… Ⅲ. ①JAVA语言－程序设计－高等学校－教材 Ⅳ. ①TP312.8

中国版本图书馆CIP数据核字(2021)第063116号

内 容 提 要

SSM 框架是目前比较主流的 Java EE 企业级框架，适用于搭建各种大型的企业级应用系统。

本书由浅入深地讲解了 SSM 框架的基础知识及应用。全书共 18 章，分为 5 个部分。第 1 部分为初识 SSM 框架（第 1 章），让读者了解 SSM 框架的概念及分工。第 2 部分为 Spring 框架（包括第 2～5 章），主要讲解 Spring 的基本知识和应用。第 3 部分为 MyBatis 框架（包括第 6～10 章），主要讲解 MyBatis 的基本知识和应用。第 4 部分为 Spring MVC 框架（包括第 11～17 章），主要讲解 Spring MVC 的相关知识。第 5 部分为 SSM 框架综合实战（第 18 章），读者可以通过案例完成对整个 SSM 框架的搭建与综合运用，并体验软件开发流程。

本书既可作为高等院校计算机相关专业的教材和辅导用书，也可作为 Java 技术的培训教材，适合广大编程爱好者阅读与使用。

◆ 主　　编　姬忠红　曹　慧　周业勤
　副 主 编　代　敏　李海斌　刘丛丛　张新健　林培光
　责任编辑　张　斌
　责任印制　王　郁　马振武
◆ 人民邮电出版社出版发行　　北京市丰台区成寿寺路 11 号
　邮编　100164　　电子邮件　315@ptpress.com.cn
　网址　https://www.ptpress.com.cn
　北京隆昌伟业印刷有限公司印刷
◆ 开本：787×1092　1/16
　印张：16.25　　2021 年 5 月第 1 版
　字数：384 千字　　2023 年 1 月北京第 6 次印刷

定价：59.80 元

读者服务热线：(010)81055256　印装质量热线：(010)81055316
反盗版热线：(010)81055315
广告经营许可证：京东市监广登字 20170147 号

前言 PREFACE

“SSM 框架开发技术”是高等学校计算机专业学生使用企业级框架进行软件设计及培养软件开发能力的重要课程，它在计算机学科的教学中起着非常重要的作用。

浪潮集团是我国领先的云计算、大数据服务商，其综合实力位列中国 IT 企业前列。山东浪潮优派科技教育有限公司（以下简称浪潮优派）是浪潮集团下属子公司，它结合浪潮集团的技术优势和丰富的项目案例，专门致力于 IT 人才的培养。本书由浪潮优派具有多年开发经验和实训经验的 Java 培训讲师撰写，通过通俗易懂的语言和丰富实用的案例，详细讲解了 Spring、Spring MVC 和 MyBatis 三大框架（以下简称“SSM”）的基本知识和应用。全书各章节知识点讲解条理清晰、循序渐进，每个知识点都配有丰富的案例演示，并用综合案例的应用演示贯穿全书知识体系。本书提供了微课视频，读者可扫描二维码直接观看。此外，除第 1 章外的各章有配套习题和上机指导，并配有教学大纲、案例源代码和电子课件等丰富的配套资源，读者可登录人邮教育社区（www.ryjiaoyu.com）下载相关资源。

全书共 18 章，具体内容如下。

第 1 章初识 SSM 框架。简要介绍了 SSM 框架及 Spring 框架、MyBatis 框架、Spring MVC 框架的概念和分工。

第 2 章 Spring 入门。介绍了 Spring 框架的体系结构及特征，Spring 的环境搭建方法，Spring IoC 容器的工作原理，以及使用 Spring IoC 机制初步完成 Service 层和 DAO 层之间的解耦。

第 3 章 Spring 的基本用法。介绍了依赖注入的概念及依赖注入的类型、配置依赖、Bean 的作用域和 Bean 的自动装配方式，以及基于注解的装配。

第 4 章面向切面编程。介绍了 AOP 的概念和相关术语，以及 AOP 的两种代理机制：JDK 动态代理、CGLIB 代理。还详细讲解了 AOP 的编程的两种方式。

第 5 章事务管理。讲解了事务的概念，事务的特性，JDBC 事务管理的步骤和操作过程，以及 Spring 事务管理的两种实现方式：编程式事务管理和声明式事务管理。

第 6 章 MyBatis 入门。对 MyBatis 的概念与优缺点、MyBatis 与 JDBC 和 Hibernate 的比较、MyBatis 的工作流程和环境配置进行了较为详细的讲解。

第 7 章基于 MyBatis 的增删改查操作。对基于 MyBatis 的增删改查操作、MyBatis 的结果类型、MyBatis 的参数类型、MyBatis 中#和$两种语法进行了讲解。

第 8 章 MyBatis 的动态 SQL 语句。介绍了 MyBatis 中的动态 SQL 元素的使用方法，还讲解了在动态 SQL 中如何使用<SQL>标签和<include>标签抽取可重用的 SQL 片段。

第 9 章使用 MyBatis 动态代理技术实现 DAO 接口。讲解了 MyBatis 中动态代理实现 DAO 接口的基本原理和具体的开发方式。

第 10 章 MyBatis 与 Spring 的整合。讲解了 MyBatis 框架与 Spring 框架整合使用的优势和必要性，以及两个框架整合的具体流程。

第 11 章 Spring MVC 入门。介绍了 Spring MVC 框架的概念、特征及核心功能，通过案例详细讲解了 Spring MVC 开发所需要的开发环境以及基本的开发步骤。

第 12 章注解式控制器开发。主要对 Spring MVC 的核心注解的应用进行了详细讲解，介绍了 Controller 和 RequestMapping 注解类型的相关知识。

第 13 章数据验证。主要介绍了声明式数据验证的用法以及验证中常见的注解方法、验证中错误消息的处理方法、统一异常处理的方法。

第 14 章拦截器。介绍了拦截器的概念及常见应用、拦截器的配置方式、多个拦截的执行流程。

第 15 章 Spring MVC 对 Ajax 的支持。介绍了 Ajax 和 JSON 的概念，以及它们在 Spring MVC 中的应用场景。还讲解了内容协商的 3 种方式。

第 16 章文件的上传和下载。主要对 Spring MVC 环境下的文件上传和文件下载进行了详细讲解。

第 17 章 SSM 框架整合。介绍了 SSM 框架的整合思路、SSM 框架的整合过程以及如何使用 SSM 框架。

第 18 章医疗信息系统。综合了前面 17 章所介绍的知识，使用 SSM 框架开发了一个项目——医疗信息系统。

本书由浪潮优派姬忠红、山东中医药大学曹慧、浪潮优派周业勤担任主编，并进行全书的审核及统稿工作；由浪潮优派代敏、李海斌、刘丛丛、张新健，山东财经大学林培光担任副主编。另外，为了使本书的内容更适合高校使用，与浪潮集团有合作关系的高校教师也参与了本书的编写工作，具体人员有山东财经大学张燕、王倩、张春云、聂秀山、邹立达，山东管理学院梁科、张婷婷，山东中医药大学马金刚，烟台大学贺鹏飞、晋刚。

由于时间仓促和编者水平有限，本书难免存在不足之处，欢迎读者朋友批评指正。

编　者

2020 年 9 月

目录 CONTENTS

01 第1章　初识SSM框架

学习目标

- 了解 SSM 框架及三个框架的分工
- 了解 Spring 框架的概念及 IoC、AOP 的概念
- 了解 Hibernate 框架的概念
- 了解 MyBatis 框架的概念
- 了解 Hibernate 框架和 MyBatis 框架的区别
- 了解 Spring MVC 框架的概念

初识 SSM 框架

SSM（Spring+Spring MVC+MyBatis）框架由 Spring、MyBatis 两个开源框架整合而成（Spring MVC 是 Spring 中的部分内容），是目前 Java 领域使用非常广泛也非常稳定的开源 Web 框架。

SSM 框架是标准的模型-视图-控制器（Model-View-Controller，MVC）模式，将整个系统划分为视图（View）层（也叫表现层）、控制器（Controller）层、业务逻辑（Service）层、数据访问（DAO）层（也叫数据持久层）。Spring、Spring MVC、MyBatis 三个框架各有分工：Spring MVC 框架负责请求的转发和视图管理，Spring 框架实现业务对象管理，MyBatis 框架则作为数据对象的持久化引擎完成数据持久化。

对 SSM 框架有了初步的认识后，本章将分别介绍 Spring、MyBatis、Spring MVC 三个框架。

1.1　Spring 框架

Spring 是于 2003 年兴起的一个轻量级的 Java 开发框架，是一个开源框架。Spring 最主要的目的就是使 Java 企业版（Java Platform Enterprise Edition，Java EE）的开发更加容易，它是为了解决企业应用开发的复杂性而创建的。同时，Spring 之所以与 Struts、Hibernate 等单层框架不同，是因为 Spring 以统一、高效的方式构造整个系统应用，并且可以将单层框架以最佳的组合糅合在一起建立一个连贯的体系。可以说 Spring 是提供了

更完善开发环境的一个框架。

Spring 使用基本的 JavaBean 来完成以前只能由企业级 JavaBean（Enterprise JavaBean，EJB）完成的事情。然而，Spring 的用途不仅仅限于服务器端的开发。从简单性、可测试性和松耦合的角度而言，任何 Java 应用都可以从 Spring 中受益。简单来说，Spring 是一个轻量级的控制反转（Inversion of Control，IoC）和面向切面编程（Aspect Oriented Programming，AOP）的容器框架。Spring 的图标如图 1-1 所示。

图 1-1　Spring 的图标

下面分别简单介绍 IoC 和 AOP，具体内容后续会详细讲解。

1. Spring IoC 简介

Spring 通过控制反转（IoC）的技术促进了松耦合。当 Spring 应用了 IoC，一个对象依赖的其他对象会通过被动的方式传递进来，而不是这个对象自己创建或者查找依赖对象。可以认为 IoC 与 Java 命名和目录接口（Java Naming and Directory Interface，JNDI）相反，不是对象从容器中查找依赖，而是容器在对象初始化时不等对象请求就主动将依赖传递给它。

2. Spring AOP 简介

Spring 提供了对面向切面编程（AOP）的丰富支持，允许通过分离应用的业务逻辑与系统级服务（例如事务管理）进行开发。应用对象只完成业务逻辑，并不负责其他的系统级关注点，例如日志或事务支持。简单地说，就是将那些与业务无关，却为业务模块所共同调用的逻辑封装起来，便于减少系统的重复代码，降低模块间的耦合度，并有利于未来的可操作性和可维护性。

Spring 的这些特征使用户能够编写更简洁、更易管理，并且更易于测试的代码。它们也为 Spring 中的各种模块提供了基础支持。

1.2 MyBatis 框架

将数据持久化保存到数据库中常用的框架有 Hibernate、MyBatis。下面分别进行介绍。

1. Hibernate 简介

Hibernate 是当前流行的对象关系映射（Object Relational Mapping，ORM）框架之一，为 Java 数据库连接（Java Database Connectivity，JDBC）提供了较为完整的封装。Hibernate 的 ORM 实现了简单的 Java 对象（Plain Ordinary Java Object，POJO）和数据库表之间的映射，以及结构化查询语言（Structured Query Language，SQL）的自动生成和执行。

2. MyBatis 简介

MyBatis 本是阿帕奇（Apache）软件基金会的一个开源项目 iBatis，2010 年这个项目迁移到了谷歌代码（Google Code），并且改名为MyBatis。MyBatis是一个基于Java的持久层框架，它支持定制化SQL、存储过程以及高级映射。MyBatis消除了绝大多数的JDBC代码和手动设置参数以及检索结果。MyBatis可以使用简单的XML或注解来配置和映射原生信息，将接口和Java的POJO映射成数据库中的记录。

3. Hibernate和MyBatis的区别

下面从两个方面简述Hibernate和MyBatis的区别。

（1）复杂程度

真正掌握Hibernate要比MyBatis困难，Hibernate比MyBatis更加重量级一些。

MyBatis框架相对简单，很容易上手。MyBatis需要我们手动编写SQL语句，回归最原始的方式，所以可以按需求指定查询的字段，提高程序的查询效率。

Hibernate也可以通过用户自己写SQL语句来指定需要查询的字段，但这样会破坏Hibernate封装以及简洁性。

（2）数据移植性

由于MyBatis所有的SQL语句都是依赖数据库书写的，所以扩展性、迁移性比较差。

Hibernate与数据库具体的关联都在可扩展标记语言（Extensible Markup Language，XML）中，所以Hibernate查询语言（Hibernate Query Language，HQL）对具体用什么数据库并不是很关心。

1.3 Spring MVC框架

Spring MVC属于Spring FrameWork的后续产品，是一个强大灵活的Web框架。Spring MVC也是一种基于Java的实现Web MVC设计模式的请求驱动（使用请求-响应模型）类型的轻量级Web框架，即使用了MVC架构模式的思想，将Web层进行层与层之间解耦。该框架的目的就是帮助我们简化开发。

本章小结

本章简要介绍了SSM框架及三个框架的分工，然后分别介绍了Spring框架、MyBatis框架、Spring MVC框架的概念。希望读者通过本章了解持久化框架Hibernate和MyBatis的区别。本章只是概要介绍三个框架，具体内容会在后面详细讲解。

习题

1. 什么是SSM框架？
2. 简述Hibernate框架和MyBatis框架的区别。

02

第2章　Spring入门

学习目标

- 了解 Spring 框架
- 了解 Spring IoC/DI 的概念
- 掌握 Spring IoC 容器的使用方法
- 掌握 Spring 框架入门案例
- 了解 Spring 5 的新特性

2.1　Spring 框架概述

2.1.1　Spring 简介

Spring 是 Java 最流行的框架之一，是一个开源的轻量级 Java SE（Java Platform Standard Edition，Java 标准版）/Java EE（Java 企业版）开发应用框架。Spring 框架由罗德·约翰逊（Rod Johnson）设计，目的是用于简化企业级应用程序的开发。

Spring 框架概述

Spring 框架除了管理对象及其依赖关系，还提供通用日志记录、性能统计、安全控制、异常处理等面向切面编程的能力，还能管理数据库事务。Spring 框架提供了一套简单的 JDBC 访问实现，还提供了与第三方数据访问框架无缝集成，如 Hibernate、Java 持久化 API（Java Persistence API，JPA），与各种 Java EE 技术整合（如 Java Mail、任务调度等）。Spring 框架提供了一套自己的 Web 层框架 Spring MVC，而且还能非常简单地与第三方 Web 框架集成。

2.1.2　Spring 的发展历程

Spring 版本的主要发展史如表 2-1 所示，读者可以简单了解一下。目前 Spring 最新的版本为 5.3。

表 2-1　Spring 版本主要发展史

发布时间	发布版本
2003 年	Spring 0.9
2004 年	Spring 1.0
2006 年	Spring 2.0 Final Released
2007 年	Spring 2.5
2011 年	Spring 3.1 GA release
2013 年	Spring 4.0
2014 年	Spring 4.1.3 released
2015 年	Spring 4.2、4.3
2016 年	Spring 4.3 GA
2017 年	Spring 5.x

2.1.3　Spring 框架的体系结构

Spring 框架体系约有 20 个模块，这些模块大致分为核心容器（Core Container）、面向切面编程（AOP）、设备支持（Instrumentation）、数据访问/集成（Data Access/Integration）、Web、报文发送（Messaging）、测试（Test）等。Spring 的体系结构如图 2-1 所示。

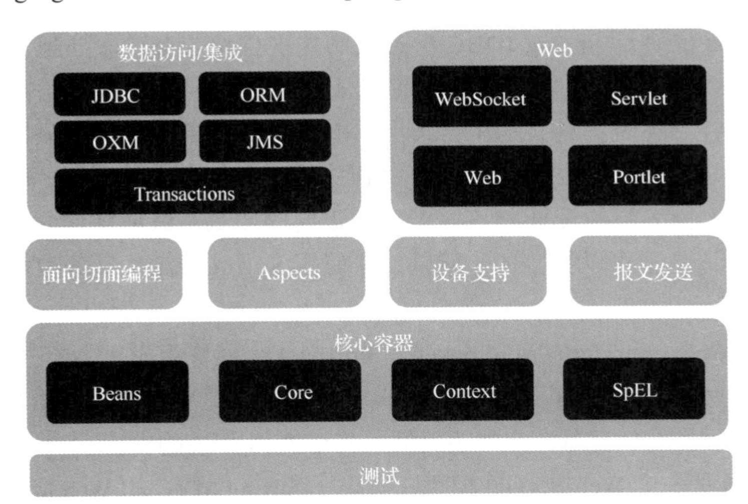

图 2-1　Spring 的体系结构

图 2-1 中主要模块的功能如下。

（1）核心容器：由 Beans、Core、Context 和 SpEL（Spring Expression Language，Spring 表达式语言）4 个模块组成。

① Beans 和 Core 模块是 Spring 框架的核心模块，包含了控制反转（IoC）和依赖注入（Dependency Injection，DI）。

② Context 模块构架于核心模块之上，它扩展了 BeanFactory（概念见 2.3 节），为其添加了 Bean（概念见 2.3 节）生命周期控制、框架事件体系以及资源加载透明化等功能。此外该模

块还提供了许多企业级支持，如邮件访问、远程访问、任务调度等。

③ SpEL 模块是统一表达式语言（EL）的扩展模块，可以查询、管理运行中的对象，同时也可以方便地调用对象方法、操作数组、集合等。它的语法类似传统 EL，但提供了额外的功能，最出色的要数函数调用和简单字符串的模板函数。这种语言的特性是基于 Spring 产品的需求而设计的，它可以非常方便地同 Spring IoC 进行交互。

（2）数据访问/集成：由 JDBC、ORM、JMS、OXM 和 Transactions 5 个模块组成。

① JDBC 模块是 Spring 提供的 JDBC 抽象框架的主要实现模块，用于简化 Spring JDBC。它主要提供 JDBC 模板方式、关系数据库对象化方式、SimpleJdbc 方式、事务管理以简化 JDBC 编程，其主要实现类包括 JdbcTemplate、SimpleJdbcTemplate 和 NamedParameterJdbcTemplate。

② ORM 模块是 ORM 框架支持的模块，主要集成 Hibernate、JPA 和 Java 数据对象（Java Data Objects，JDO）用于资源管理、数据访问对象（Data Access Object，DAO）的实现和事务策略。

③ JMS（Java Messaging Service，Java 消息服务）模块提供 Java 消息传递服务，能够发送和接收信息，自 Spring Framework 4.1 以后，它还提供了对 Spring-messaging 模块的支撑。

④ OXM（Object-XML Mapping，对象-XML 映射）模块主要提供一个抽象层以支撑 OXM（它可将 Java 对象映射成 XML 数据，或者将 XML 数据映射成 Java 对象），如 JAXB、Castor、XMLBeans、JiBX 和 XStream 等。

⑤ Transactions 模块是 Spring JDBC 事务控制实现模块。Spring 框架对事务做了很好的封装，通过它的 AOP 配置，可以灵活地配置在任何一层。但是在很多的需求和应用中，直接使用 JDBC 事务控制还是有其优势的。其实，事务是以业务逻辑为基础的，一个完整的业务应该对应业务层里的一个方法，如果业务操作失败，则整个事务回滚。所以，事务控制是应该放在业务层的。持久层的设计应该遵循一个重要的原则：保证操作的原子性，即持久层里的每个方法都应该是不可以分割的。所以，在使用 Spring JDBC 事务控制时，应注意其特殊性。

（3）Web：由 Web、Servlet、WebSocket 和 Portlet 4 个模块组成。

① Web 模块为 Spring 提供了最基础的 Web 支持，主要建立在核心容器之上，通过 Servlet 或者 Listeners 来初始化 IoC 容器，此外还提供一些与 Web 相关的支持。

② Servlet 模块是一个 Web-Servlet 模块，实现了 Spring MVC 的 Web 应用。

③ WebSocket 模块是 Spring 4.0 新增的模块，它提供了 WebSocket 和 SockJS（一个浏览器上运行的 JavaScript 库）。

④ Portlet 模块提供了在 Portlet 环境中使用 MVC 实现，类似 Servlet 模块的功能。

（4）其他模块：Spring 的其他模块还有 AOP、Aspects、Instrumentation、Messaging、Test 等。

① AOP 模块是 Spring 的另一核心模块，它提供了面向切面编程的实现。AOP 作为继面向对象编程后，对程序员影响最大的编程思想之一，极大地开拓了程序员对于编程的思路。在 Spring 中，它以 JVM（Java Virtual Machine，Java 虚拟机）的动态代理技术为基础，设计出了一系列的 AOP 横切实现，如前置通知、返回通知、异常通知等，同时，Pointcut 接

口来匹配切入点，可以使用现有的切入点来设计横切面，也可以扩展相关方法根据需求进行切入。

② Aspects 模块集成自 AspectJ 框架，主要为 Spring AOP 提供多种 AOP 实现方法。

③ Instrumentation 模块是基于 Java SE 中的 java.lang.instrument 进行设计的，应该算是 AOP 的一个支援模块，其主要作用是在 JVM 启用时，生成一个代理类，程序员通过代理类在运行时修改类的字节，从而改变一个类的功能，实现 AOP 的功能。

④ Messaging 模块是从 Spring 4.0 开始新加入的一个模块，主要职责是为 Spring 框架集成一些基础的报文传送应用。

⑤ Test 模块主要为单元测试和集成测试提供支持，毕竟在不需要发布（程序）到应用服务器或者连接到其他企业设施的情况下，就能够执行一些集成测试或其他测试对于任何企业都是非常重要的。

2.1.4 Spring 的特征

Spring 具有以下基本特征。

（1）轻量级

从大小和系统开支上说，Spring 都是轻量级的，而且 Spring 是非侵入式的，基于 Spring 开发的系统一般都不依赖于 Spring 的类。

（2）反向控制

Spring 提倡反向控制实现松耦合。使用 IoC，对象是被动接受依赖类而不是自动查找，Spring 容器会在实例化类的时候主动把该类所需的依赖对象注入该类。

（3）面向切面

Spring 为面向切面编程提供了强大的支持，把业务逻辑从系统服务中分离了出来，系统对象只做它们该做的事情，而不需要关心其他事情。

（4）方便程序的测试

Spring 可以用非容器依赖的编程方式进行绝大多数的测试工作。在 Spring 里，测试不再是复杂的操作，而是随手可做的事情。例如，Spring 支持 JUnit，JUnit 可以通过注解方便地测试 Spring 程序。

（5）集成各种优秀框架

Spring 可以集成各种优秀的开源框架，如 Struts、Hibernate、Hessian、Quartz 等。

（6）对声明式事务的支持

在 Spring 中，我们可以从单调的事务管理代码中解脱出来，通过声明方式灵活地进行事务的管理，这样我们可以专注于业务逻辑开发，提高开发效率和质量。

2.2 Spring IoC/DI 概述

Spring IoC/DI 概述

下面介绍 Spring 控制反转（IoC）和依赖注入（DI）的概念。

1. IoC 的概念

IoC 是 Spring 的核心概念，控制反转就是应用本身不负责依赖对象的创建及维护，依赖对象的创建及维护是由外部容器 Spring 负责的。这样控制权就由应用程序转移到了外部容器 Spring，控制权的转移就是所谓的反转。

当某个 Java 实例（调用者）需要另一个 Java 实例（被调用者）时，在传统的程序设计过程中，通常由调用者来创建被调用者的实例，而控制反转就转移了控制权，即把调用者的创建转移到了外部容器 Spring。

控制反转显然是一个抽象的概念，下面举一个生活中的例子来简单说明一下。

在以前没有饮品店的时代，我们想喝新鲜橙汁，最直接的做法就是：买榨汁机、橙子，然后准备水，自己去榨橙汁。这些都是我们自己“主动”创造橙汁的过程，如图 2-2 所示。

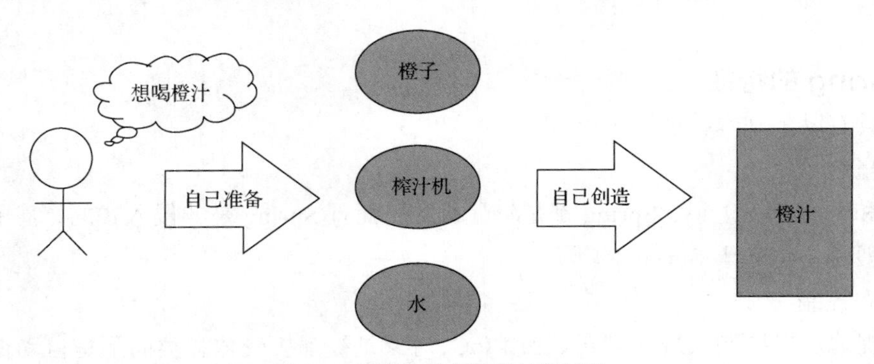

图 2-2 我们主动创造橙汁的过程

然而到了现在，由于饮品店的盛行，当我们想喝橙汁时，第一个想法就变成了找到饮品店的联系方式，通过网络、电话等渠道联系饮品店并描述需要、地址等，然后下订单等待，过一会儿就会有人送来橙汁。图 2-3 是饮品店榨好果汁，然后给我们配送橙汁的过程。其中饮品店就相当于 Spring 容器，橙汁就相当于被调用者 Java 实例，我们就相当于调用者 Java 实例。

图 2-3 我们被动接受橙汁的过程

2. DI 的概念

IoC 也是一种设计模式，Spring 框架采用依赖注入（DI）实现 IoC。

依赖注入是指在运行期，由外部容器 Spring 动态地将依赖对象注入组件中。即通过使用 Spring 框架，开发人员将不必在自己的代码中维护对象之间的依赖关系，只需在配置文件中或使用注解对 Bean 之间的依赖关系进行设定，Spring 容器会自动依据配置信息或注解来维护对象之间的依赖关系。

2.3　Spring IoC 容器

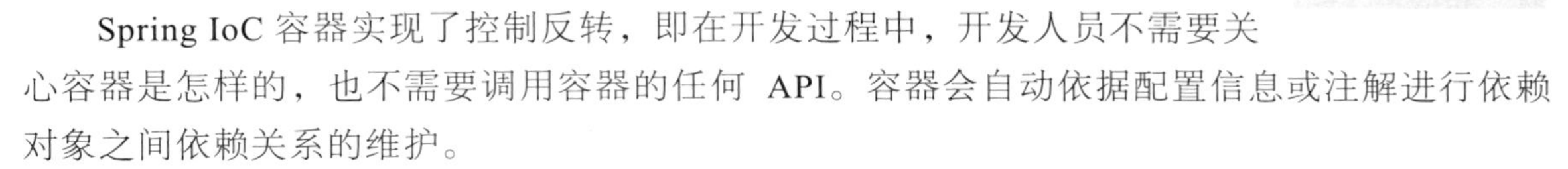

1. Spring IoC 容器概述

Spring IoC 容器实现了控制反转，即在开发过程中，开发人员不需要关心容器是怎样的，也不需要调用容器的任何 API。容器会自动依据配置信息或注解进行依赖对象之间依赖关系的维护。

在 Spring 中，最重要的两个包是 org.springframework.beans 包和 org.springframework.context 包，提供了 Spring IoC 容器的基本功能，这两个包是 Spring IoC 容器的基础。

Spring IoC 容器中有两个非常重要的接口：BeanFactory 和 ApplicationContext。

（1）BeanFactory 接口提供了一种能够管理任何类型对象的高级配置机制。Spring 框架中的核心接口是 BeanFactory 接口，它是工厂模式的具体实现。BeanFactory 使用控制反转对应用程序的配置和依赖性规范与实际的应用程序代码进行了分离。但 BeanFactory 实例化后并不会自动实例化 Bean，只有当 Bean 被使用时，BeanFactory 才会对该 Bean 进行实例化与依赖关系的配置。

（2）ApplicationContext 是 BeanFactory 的子接口。ApplicationContext 与 BeanFactory 不同，ApplicationContext 实例化后会自动对所有的单实例 Bean 进行实例化与依赖关系的配置，使之处于待用状态。它能更容易地集成 Spring 的 AOP 功能、消息资源处理（如在国际化中使用）、事件发布和特定的上下文应用层（如在网站应用中的 WebApplicationContext）。

总之，BeanFactory 提供了配置框架和基本方法，ApplicationContext 添加了更多企业特定的功能。

在 Spring 中，由 Spring IoC 容器管理的对象叫作 Bean。Bean 是由 Spring IoC 容器实例化、组装和以其他方式管理的对象。此外，Bean 只是应用中许多对象中的一个。Bean 以及它们之间的依赖关系通过容器配置元数据反映出来。

2. Spring IoC 容器的工作原理

Spring IoC 容器负责实例化、配置和组装 Bean，这里的 Bean 是一种广泛意义上的 JavaBean 对象。

Spring IoC 容器通过读取配置元数据获取有关要实例化、配置和组装的对象的指令。配置元数据以 XML、Java 注解或 Java 代码表示。它允许我们表达组成应用程序的对象以及这些对象之间丰富的相互依赖性。

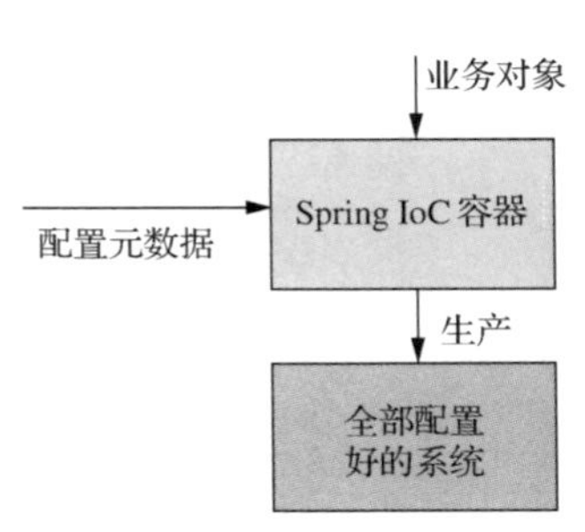

图 2-4　Spring IoC 容器

从图 2-4 可以看到，我们可以在 Spring IoC 容器中配置应用程序需要的信息，如 POJO 业务类、这些业务类（Bean）之间的相互依赖。Spring 容器负责实例化、定位、配置应用程序中的对象以及建立这些对象间的依赖。整个 Spring IoC 容器就是一个大的工厂，为应用程序提供 Bean。

3. Spring IoC 容器的核心接口

ApplicationContext 接口代表 Spring IoC 容器，是核心接口。Spring 提供了多种常用的 ApplicationContext 接口的实现类。在独立的应用程序中通常会创建一个 ClassPathXmlApplicationContext 或 FileSystemXmlApplicationContext 实现类（见表 2-2）。Spring IoC 容器通过读取 Bean 的配置完成创建 Bean 工厂，我们常用 XML 的形式配置元数据，也可以使用 Java 注解完成 Bean 的配置，但要通过提供少量 XML 配置来声明启用对注解的支持。

表 2-2 ApplicationContext 实现类

ApplicationContext 实现类	描述
org.springframework.context.support.ClassPathXmlApplicationContext	从类路径中的 XML 文件载入上下文定义信息
org.springframework.context.support.FileSystemXmlApplicationContext	从文件系统中的 XML 文件载入上下文定义信息

既然ClassPathXmlApplicationContext和FileSystemXmlApplicationContext都是ApplicationContext接口的实现类，那它们之间有什么区别呢？

ClassPathXmlApplicationContext 从类路径载入应用上下文，并且将 Bean 载入 BeanFactory。而 FileSystemXmlApplicationContext 从文件系统载入应用上下文，并且将 Bean 载入 BeanFactory。因为 FileSystemXmlApplicationContext 从外部文件中配置 Bean，对于软件的移植性不好，所以项目中常用 ClassPathXmlApplicationContext 加载 Bean 的配置。

样例代码如下：

```
ApplicationContext context1 = new FileSystemXmlApplicationContext("C:/beans.xml");
MyBean myBean = context1.getBean("myBean");

ApplicationContext context = new ClassPathXmlApplicationContext("beans.xml");
MyBean myBean = context.getBean("myBean");
```

4. 配置元数据

如图 2-4 所示，Spring IoC 容器使用配置元数据形式完成实例化、配置和组装对象。

传统上，配置元数据以简单直观的 XML 格式提供，本章的大部分内容用于传达 Spring IoC 容器的关键概念和功能。

以下示例显示了基于 XML 的配置元数据的基本结构。

```
<?xml version="1.0" encoding="UTF-8"?>
<beans xmlns="http://www.springframework.org/schema/beans"
    xmlns:xsi="http://www.w3.org/2001/XMLSchema-instance"
    xsi:schemaLocation="http://www.springframework.org/schema/beans
        http://www.springframework.org/schema/beans/spring-beans.xsd">
    <bean id="..." class="...">
    </bean>
</beans>
```

其中 bean 元素是用来配置 JavaBean 对象的，该 id 属性是一个字符串，用于标识单个

JavaBean 的定义。该 class 属性定义 JavaBean 的类型并使用完全限定的类名。

实例化 Spring IoC 容器非常简单。提供给 ApplicationContext 构造函数的位置路径实际上是资源字符串，允许容器从各种外部资源（如本地文件系统等）加载配置元数据。

实例化 Spring IoC 容器样例代码如下：

```
ApplicationContext context = new ClassPathXmlApplicationContext("beans.xml");
```

2.4　Spring 框架入门案例

2.4.1　环境准备

环境准备

Java 开发需要一套环境，下面进行简单介绍。

1. JDK

Java 开发工具包（Java Development Kit，JDK）提供了 Java 开发环境和运行环境，是所有 Java 应用程序的基础。JDK 包括一组 API 和 JRE，这些 API 是构建 Java 应用程序的基础，而 JRE 是运行 Java 应用程序的基础。JDK 的版本需要在 8.0 以上，本书用 JDK 8.0 版本进行讲解。读者可以从 Oracle 官网下载 JDK 安装文件。

2. Eclipse 集成开发工具

为了提高程序的开发效率，我们使用集成开发工具（Integrated Development Environment，IDE）进行 Java 程序开发。正所谓“工欲善其事，必先利其器”，接下来介绍一种 Java 常用的开发工具——Eclipse。Eclipse 是由 IBM 公司开发的一款功能完整且成熟的集成开发环境，它是一个开源的、基于 Java 的可扩展开发平台，是目前流行的 Java 语言开发工具。Eclipse 具有强大的代码编排功能，可以帮助程序开发人员完成语法修正、代码修正、补全文字、信息提示等编码工作，大大提高了程序开发的效率。读者可以从 Eclipse 官网下载其安装包。

3. Tomcat 服务器

Tomcat 是由 Apache 开发的一个 Servlet 容器，因为 Tomcat 技术先进、性能稳定，而且免费，故而深受 Java 爱好者的喜爱并得到了部分软件开发商的认可，成为目前比较流行的 Web 应用服务器。Tomcat 服务器是一个免费的开放源代码的 Web 应用服务器，属于轻量级应用服务器。本书用 Tomcat 8.0 版本进行讲解。Tomcat 的安装文件可以从其官网下载。

2.4.2　编写 Spring 框架入门案例

编写 Spring 框架入门案例

1. 准备 Spring 开发包及 logging 日志包

（1）从 Spring 官网下载 Spring 开发包，如图 2-5 所示。

开发包解压后的目录结构如图 2-6 所示。

docs 包：该包中存放了 API 和规范。

libs 包：该包中存放了支持 Spring 框架的所有 JAR 包。

schema 包：该包中存放了 Spring 中 XML 文档的 schema 约束。

（2）下载 commons-logging-1.2.jar，如图 2-7 所示。

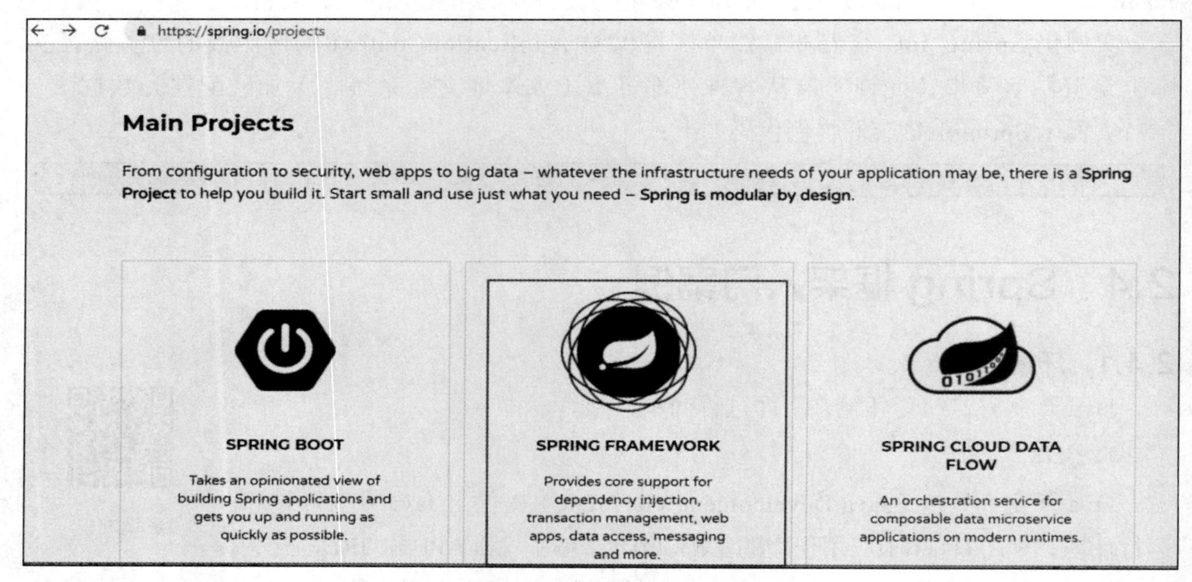

图 2-5　下载 Spring 开发包

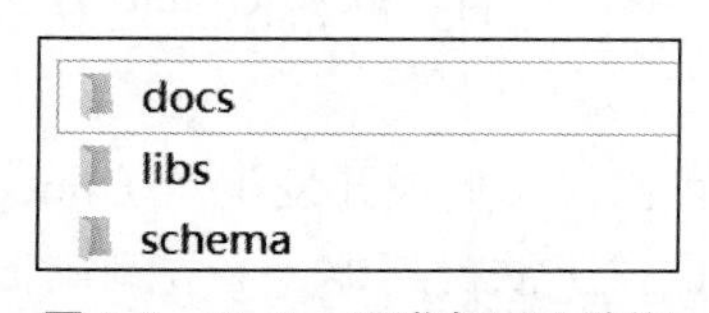

图 2-6　Spring 开发包目录结构

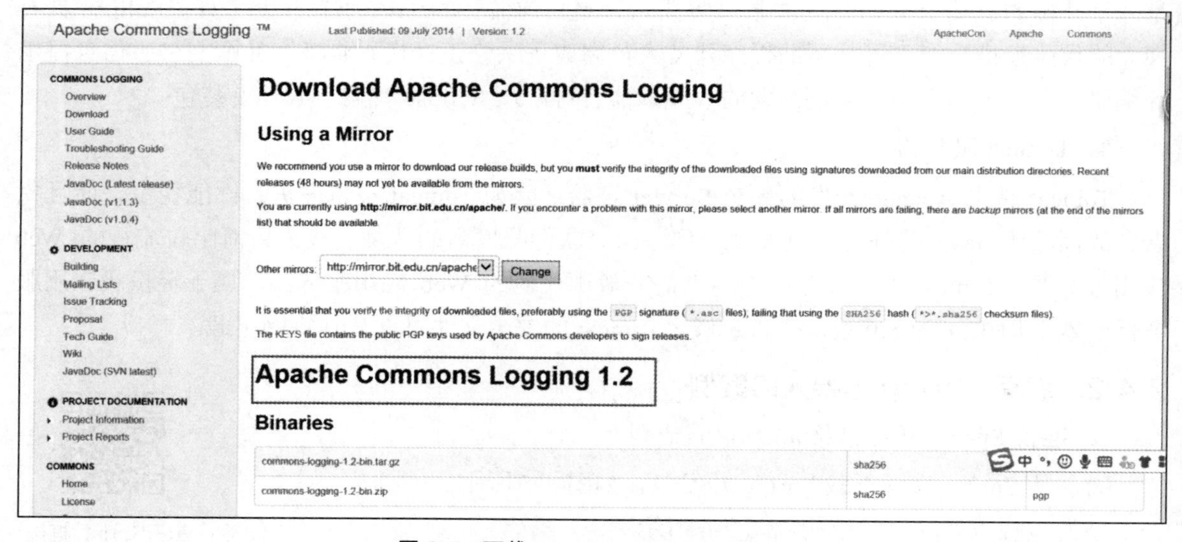

图 2-7　下载 commons-logging-1.2.jar

2．建立 Java 工程并添加 JAR 包到编译路径上

打开 Eclipse 建立 Java Project，这里只测试 Spring 入门，没有实际的视图层，可以不用建立 Web Project，建立 Java Project 即可。读者可以先了解项目工程目录结构，如图 2-8 所示。

```
Spring5_Ch01_FirstDemo
  src
    com.inspur.dao
      UserDao.java
    com.inspur.dao.impl
      UserDaoImpl.java
    com.inspur.service
      UserService.java
    com.inspur.service.impl
      UserServiceImpl.java
    com.inspur.ui
      Client.java
    beans.xml
  JRE System Library [jdk1.8.0_191]
  Referenced Libraries
  lib
```

图 2-8　Spring 入门案例目录

我们需要添加 JAR 包到项目的编译路径上。在项目工程中建立 lib 文件夹，对 spring-framework-5.0.2.RELEASE-dist.zip 解压后，将 libs 文件夹中的 spring-beans-5.0.2.RELEASE.jar、spring-context-5.0.2.RELEASE.jar、spring-core-5.0.2.RELEASE.jar、spring-expression-5.0.2.RELEASE.jar 复制到 lib 文件夹，将 commons-logging-1.2.jar 也复制到 lib 文件夹下，然后将这 5 个 JAR 包放到编译路径上。

3. 实例操作

【实例 2.1】模拟完成保存用户信息的操作。具体源代码如下所示（代码详见 Spring5_Ch02_FirstDemo\src\com\inspur\dao\UserDao.java；Spring5_Ch02_FirstDemo\src\com\inspur\dao\impl\UserDaoImpl.java；Spring5_Ch02_FirstDemo\src\com\inspur\service\UserService.java；Spring5_Ch02_FirstDemo\src\com\inspur\service\impl\UserServiceImpl.java；Spring5_Ch02_FirstDemo\src\com\inspur\ui\Client.java；Spring5_Ch02_FirstDemo\src\com\inspur\beans.xml）。

（1）UserDao.java

```
package com.inspur.dao;
public interface UserDao {
  public int saveUser();
}
```

（2）UserDaoImpl.java

```
package com.inspur.dao.impl;
import com.inspur.dao.UserDao;
public class UserDaoImpl implements UserDao{
  public int saveUser() {
      //这里只是模拟，没有连接数据库进行保存操作
      return 0;
  }
}
```

（3）UserService.java

```
package com.inspur.service;
public interface UserService {
```

```
    public int saveUser();
}
```

（4）UserServiceImpl.java

```
package com.inspur.service.impl;
import com.inspur.dao.UserDao;
import com.inspur.dao.impl.UserDaoImpl;
import com.inspur.service.UserService;
public class UserServiceImpl implements UserService{
  private UserDao userDao=new UserDaoImpl();
  public UserServiceImpl() {
       System.out.println("UserServiceImpl 对象创建了");
  }
  public int saveUser() {
       int count=userDao.saveUser();
       System.out.println("保存完毕");
       return count;
  }
}
```

（5）beans.xml

```
<?xml version="1.0" encoding="UTF-8"?>
<beans xmlns="http://www.springframework.org/schema/beans"
     xmlns:xsi="http://www.w3.org/2001/XMLSchema-instance"
  xsi:schemaLocation="http://www.springframework.org/schema/beans
   http://www.springframework.org/schema/beans/spring-beans.xsd">
     <bean id="userService" class="com.inspur.service.impl.UserServiceImpl">
     </bean>
     <bean id="userDao" class="com.inspur.dao.impl.UserDaoImpl">
     </bean>
</beans>
```

（6）Client.java

```
package com.inspur.ui;
import org.springframework.context.ApplicationContext;
import org.springframework.context.support.ClassPathXml
ApplicationContext;
import com.inspur.dao.UserDao;
import com.inspur.service.UserService;
public class Client {
     public static void main(String[] args) {
          ApplicationContext beanFactory=new ClassPathXmlApplication
     Context("beans.xml");
          UserService userService=beanFactory.getBean("userService",
     UserService.class);
          UserDao userDao=(UserDao) beanFactory.getBean("userDao");
          System.out.println(userDao);
          System.out.println(userService);
     }
}
```

选中 Client.java 文件，单击鼠标右键，选择“Run As”菜单中的子菜单“1 Java Application”，运行后，在控制台上显示的结果如图 2-9 所示。

```
UserServiceImpl对象创建了
com.inspur.dao.impl.UserDaoImpl@4c70fda8
com.inspur.service.impl.UserServiceImpl@224edc67
```

图 2-9 实例 2.1 的运行结果

【程序解析】

在 UserServiceImpl.java 中，代码 private UserDao userDao =new UserDaoImpl()中的 userDao 是 UserServiceImpl 的依赖,因依赖注入将在后续章节讲解,故而这里依赖对象 userDao 先用 new 操作符完成。

在 Client.java 中，代码“ApplicationContext beanFactory=new ClassPathXmlApplicationContext ("beans.xml");”为初始化 Spring IoC 容器（Spring 容器），Spring 容器会创建好所有的 Bean 对象，然后根据 Bean 的 id 从 Spring 容器中取出 JavaBean 对象。得到 JavaBean 对象有以下两种方式。

第一种方式：使用 ApplicationContext 接口的父接口 BeanFactory 中的<T> T getBean(String name,@Nullable Class<T> requiredType)方法，该方法中第一个参数为 Bean 的 id，第二个参数为要转化的 JavaBean 字节码文件。样例代码为“UserService userService=beanFactory.getBean ("userService",UserService.class);”。

第二种方式：使用 Object getBean(String name)方法，该方法参数为 Bean 的 id。使用该方法得到 Object 返回值类型，然后强制类型转化为所需的 JavaBean 类型。

图 2-9 中显示 UserDaoImpl 对象和 UserServiceImpl 对象已经从 Spring 容器中得到了。

2.5 Spring 5 的特性

Spring 5 的特性

Spring 5 有哪些特性呢？下面进行简单介绍。

1. JDK 版本升级

Spring 5 的代码基于 Java 8 的语法规范，因此要想使用 Spring 5，JDK 的版本至少要在 8.0 以上。

2. Core 框架修订

基于 Java 8 的反射增强，方法的参数在 Spring 5 中可以高效地访问核心的 Spring 接口，提供利用了 Java 8 的默认接口实现，default 方法提供了一些可选的声明@Nullable 和@NotNull 注解，精确标记了方法的参数和返回值，这样可以在编译的时候处理 null 值，而不至于在运行的时候抛出空指针异常 NullPointerExceptions。

在日志端，Spring 5 提供了 Common Logging 的桥接模块 spring-jcl，代替了标准的 Common Logging，同时它还可以自动检测 Log4J 2.x、SLF4J、JUL(java.util.logging)，而不需要额外的依赖。

3. Kotlin 函数式编程

Spring 5 引入了 JetBrains 的 Kotlin 语言支持，Kotlin 是一种支持函数式编程的面向对象编程语言，也运行在 JVM 之上。有了 Kotlin 的支持，开发者可以使用 Spring 的函数式编程处理 Web 的入口点和 Bean 的注册。

例如，可以写成如下的代码风格。

在 Web 的入口点的时候：

```
{
    ("/movie" and accept(TEXT_HTML)).nest {
        GET("/", movieHandler::findAllView)
        GET("/{card}", movieHandler::findOneView)
    }
    ("/api/movie" and accept(APPLICATION_JSON)).nest {
        GET("/", movieApiHandler::findAll)
        GET("/{id}", movieApiHandler::findOne)
    }
}
```

在注册 Bean 的时候：

```
val context = GenericApplicationContext {
    registerBean()
    registerBean { Cinema(it.getBean()) }
}
```

4. 响应式编程模型

Spring 5 的响应式 Web 编程 Reactive Streams（响应式流），是 NetFlix、Pivotal、Typesafe、Red Hat、Oracle、Twitter 和 Spray.io 共同开发的一套规范。它提供了一些通用的 API，实现自己控制，就像 Hibernate 的 JPA，JPA 是 API，Hibernate 是实现。

响应式流是 Java 9 的正式模块，但是在 Java 8 中，需要引入额外的依赖。Spring 5 的流式支持基于响应式流的 API 的 Project Reactor。

Spring 5 中有一个新模块叫作 spring-webflux，可以支持响应式的 HTTP 和 WebSocket 客户端。

通过 spring-webflux 可以创建 Webclient，它是响应式和非阻塞的 RestTemplate 的替代。代码示范如下。

```
WebClient webClient = WebClient.create();
Mono person = webClient.get()
.uri("http://localhost:8080/movie/42")
.accept(MediaType.APPLICATION_JSON)
.exchange()
.then(response -> response.bodyToMono(Movie.class));
```

5. 测试改进

Spring 5 完全支持 JUnit 5，在 TestContext 框架中，可以并行地执行测试。对于响应式的编程，Spring-test 提供了 WebTestClient 来测试 spring-webflux。当然，Spring 5 还是支持 JUnit 4 的，在未来的一段时间里，JUnit 4 都是会存在的。

6. 额外库支持

目前 Spring 5 支持下面的库版本：Jackson 2.6+、Ehcache 2.10+/3.0 GA、Hibernate 5.0+、JDBC

4.0+、XMLUnit 2.x+、OkHttp 3.x+、Netty 4.1+。

在 API 层面上，Spring 5 不再支持如下的包：beans.factory.access、jdbc.support.nativejdbc、spring-aspects 模块的 mock.staticmock、web.view.tiles2M（Tiles 3 是最低要求的版本）、orm.hibernate 3 和 orm.hibernate 4。不再支持如下的库：Portlet、Velocity、JasperReports、XMLBeans、JDO、Guava。

本章小结

本章介绍了 Spring 框架的体系结构及特征，读者需要掌握 Spring IoC/DI、Spring IoC 容器的概念。了解以上基本概念后，读者可通过 Spring 框架入门案例掌握 Spring 的环境搭建方法，理解 Spring IoC 容器的工作原理，掌握从 Spring IoC 容器中根据 Bean 的 id 得到 JavaBean 对象的两种方式。通过第一个入门案例，读者可以初步体验使用 Spring IoC 机制初步完成 Service 层和 DAO 层之间的解耦。

习题

1. 简述 Spring 的框架及其特征。
2. 简述 Spring IoC/DI 的概念。

上机指导

开发一个 Java 工程，引入 Spring 框架，完成模拟用户登录功能。

03 第3章　Spring的基本用法

学习目标

- 掌握依赖注入的类型
- 掌握依赖配置的方式
- 掌握 Bean 的作用域范围
- 掌握 Bean 的自动装配方法
- 掌握 Bean 的基于 Annotation 的装配方法

3.1　依赖注入

典型的企业级应用程序不只包含单个对象（或 Spring 用法中的 Bean），即使是最简单的应用程序也需要多个 Bean 对象协作完成。Bean 对象之间的关系就叫作依赖。本节我们将介绍如何定义多个独立的 Bean，以及定义 Bean 之间的依赖关系。

Spring 框架的核心功能之一就是通过依赖注入的方式来管理 Bean 之间的依赖关系。

关于依赖注入的概念请参见 2.2 节。

3.2　依赖注入的类型

依赖注入（DI）是一个过程，通过这个过程，Spring IoC 容器创建对象时会动态地将其依赖的对象注入 Bean 对象中。

依赖注入的类型

Spring 容器支持多种形式的 Bean 的配置，下面讲解基于 XML 的配置方式。基于 XML 的配置所支持的依赖注入的类型常用的有两种，一种是基于构造函数的依赖注入，另一种是基于 setter 的依赖注入。

3.2.1 Bean 的配置

作为 Spring 核心机制的依赖注入，改变了传统的编程习惯，对组件的实例化不再由应用程序完成，转而交由 Spring 容器完成，在需要时注入应用程序中，从而对组件之间的依赖关系进行了解耦。这一切都离不开 Spring 配置文件中使用的 bean 元素。也就是 Spring 容器需要根据 Bean 的配置创建 Bean 对象及配置 Bean 之间的依赖关系。Spring 容器支持 XML 和 Properties 两种格式的配置文件，在实际项目开发中最常用的是 XML 格式配置方式。

在 Spring 中，XML 配置文件的根标签为<beans>，<beans>中包含了多个<bean>子标签，每个<bean>配置一个 Bean 对象。

定义 Bean 的示例代码如下所示。

```
    <beans xmlns="http://www.springframework.org/schema/beans"
    xmlns:xsi="http://www.w3.org/2001/XMLSchema-instance"
xsi:schemaLocation="http://www.springframework.org/schema/beans
        http://www.springframework.org/schema/beans/spring-beans.xsd">
       <bean id="person1" class="com.inspur.Person1"></bean>
       <bean name="person2" class="com.inspur.Person2"></bean>
    </beans>
```

在上述代码中，分别使用 id 和 name 属性定义了两个 Bean，并使用 class 属性指定了 Bean 对应的实现类的全路径。

<bean>标签中可以包含多个子标签和属性，其常用的属性及子标签如表 3-1 所示。

表 3-1　<bean>标签的常用属性及子标签

属性名称	描述
id	id 是一个 Bean 的唯一标识符，Spring 容器对 Bean 的配置和管理都可通过该属性完成
name	Spring 容器同样可以通过此属性对容器中的 Bean 进行配置和管理，name 属性中可以为 Bean 指定多个名称，每个名称之间用逗号或分号隔开
class	该属性指定了 Bean 的具体实现类，它必须是一个完整的类名，使用类的完整路径
scope	用于设定 Bean 实例的作用域，其属性值有 singleton（单例）、prototype（原型）、request、session 和 global Session。默认值是 singleton
constructor-arg	<bean>标签的子标签，可以使用此标签传入构造参数进行实例化。该标签的 index 属性可指定构造参数的序号（从 0 开始），type 属性可指定构造参数的类型
property	<bean>标签的子标签，用于调用 Bean 实例中的 set 方法完成属性赋值，从而完成依赖注入。该标签的 name 属性可指定 Bean 实例中的相应属性名
ref	<property>和<constructor-arg>等标签的子标签，该属性可指定对 Bean 工厂中某个 Bean 实例的引用
value	<property>和<constructor-arg>等标签的子标签，用于直接指定一个常量值
list	用于封装 List 或数组类型的依赖注入
set	用于封装 Set 类型属性的依赖注入
map	用于封装 Map 类型属性的依赖注入
entry	<map>标签的子标签，用于设置一个键值对。其 key 属性可指定字符串类型的键值，ref 或 value 子标签可指定其值

3.2.2 基于构造函数的依赖注入

基于构造函数的依赖注入由 Spring IoC 容器调用具有多个参数的构造函数来完成，每个参数均表示一个依赖项。

基于构造函数的依赖注入

1. 构造函数参数解析

通过使用参数的类型进行构造函数参数解析匹配。如果 bean 定义的构造函数参数中不存在潜在的歧义，那么在 bean 定义中定义构造函数参数的顺序就是在实例化 Bean 时将这些参数提供给适当的构造函数的顺序。

以下代码示例显示了一个只能通过构造函数注入进行依赖注入的类，且构造器的参数为其他的 Bean 对象。

```
package com.inspur;
public class BeanOne {
     private BeanTwo beanTwo;
     private BeanThree beanThree;
     public BeanOne(BeanTwo beanTwo, BeanThree beanThree) {
          this.beanTwo=beanTwo;
          this.beanThree=beanThree;
     }
}
```

其中，BeanTwo、BeanThree 是其他的两个 JavaBean 类。这里不展开编写 BeanTwo 和 BeanThree 类了，仅代表参数为其他 JavaBean 的情况。

2. **构造函数参数类型匹配**

关于构造函数参数类型匹配有以下几种方式。

（1）默认根据构造器参数列表顺序进行匹配

假设 BeanTwo 和 BeanThree 类与继承无关，则不存在类型匹配上的歧义。因此，我们配置 Bean 对象时，不需要在<constructor-arg/>中显式指定构造函数参数索引或类型。因为 BeanOne 构造器中两个参数都是自定义的 Java 类，所以构造器参数的值用 ref 指定的 bean 的 id。在 JavaBean 的配置文件中配置代码如下。

```
<beans>
   <bean id="beanTwo" class="com.inspur.BeanTwo"/>
   <bean id="beanThree" class="com.inspur.BeanThree"/>
   <bean id="beanOne" class="com.inspur.BeanOne ">
       <constructor-arg ref="beanTwo"/>
       <constructor-arg ref="beanThree"/>
   </bean>
</beans>
```

（2）使用构造函数参数类型进行匹配

JavaBean 的构造器的参数使用简单类型时，如 int、double、String 等，则在 bean 的配置文件中，可使用<value>true</value>进行配置。这时 Spring 无法确定值的类型，因此无法在没有帮助的情况下按类型进行匹配。

以下代码显示了构造器的参数为简单类型的 Bean 类。

```
package com.inspur;
public class ExampleBean {
    private int years;
    private String answer;
    public ExampleBean(int years, String answer) {
         this.years = years;
         this.answer = answer;
    }
}
```

对于构造器的参数为简单类型的场景，如 int、String 等，可以使用 type 属性显式指定构造函数参数的类型，则容器可以根据 type 指定的类型完成和 Bean 对象中构造器参数的匹配。在 JavaBean 的配置文件中配置代码如下。

```
<bean id="exampleBean" class="com.inspur.ExampleBean">
    <constructor-arg type="int" value="2019"/>
    <constructor-arg type="java.lang.String" value="yes"/>
</bean>
```

（3）使用构造函数参数索引进行匹配

对于构造函数参数的配置，还可以使用 index 属性显式指定构造函数参数的索引，索引位置从 0 开始。在 JavaBean 的配置文件中配置代码如下。

```
<bean id="exampleBean" class="com.inspur.ExampleBean">
    <constructor-arg index="0" value="2019"/>
    <constructor-arg index="1" value="yes"/>
</bean>
```

使用参数索引进行匹配，除了解决多个参数为简单值的歧义外，指定索引还可以解决构造函数具有相同类型的两个参数的歧义。

（4）使用构造函数参数名称进行匹配

使用构造函数参数名称进行匹配，可以消除因构造函数中参数类型相同而产生的歧义，在 JavaBean 的配置文件中配置代码如下。

```
<bean id="exampleBean" class="com.inspur.ExampleBean ">
    <constructor-arg name="years" value="2019"/>
    <constructor-arg name="answer" value="yes"/>
</bean>
```

3.2.3　基于 setter 的依赖注入

在调用无参数构造函数实例化 Bean 之后，基于 setter 的依赖注入由 Bean 上的容器调用 setter 方法完成。

以下代码显示了一个只能通过 setter 方法进行依赖注入的类。

```
package com.inspur;
public class BeanOne {
     private BeanTwo beanTwo;
     private BeanThree beanThree;
     public void setBeanTwo(BeanTwo beanTwo) {
         this.beanTwo = beanTwo;
     }
     public void setBeanThree(BeanThree beanThree) {
         this.beanThree = beanThree;
     }
}
```

对应以上示例，在 JavaBean 的配置文件中使用 property 元素配置属性，具体配置代码如下。

```
<beans>
   <bean id="beanTwo" class="com.inspur.BeanTwo"/>
   <bean id="beanThree" class="com.inspur.BeanThree"/>
   <bean id="beanOne" class="com.inspur.BeanOne ">
```

```
        <property name="beanTwo" ref="beanTwo"/>
        <property name="beanThree" ref="beanThree"/>
    </bean>
</beans>
```

根据属性类型为简单数据类型、数组类型、集合类型等，property 元素有不同的配置方式。我们在后续案例中将会具体讲解。

【实例 3.1】 验证构造器依赖注入。

具体源代码如下所示（代码详见 Spring5_Ch03_inject01\src\com\inspur\entity\User.java；Spring5_Ch03_inject01\src\beans.xml；Spring5_Ch03_inject01\src\com\inspur\test\Test.java）。

（1）User.java

```
package com.inspur.entity;
import java.util.Date;
public class User {
        private int userId;
        private String name;
        private Date birth;
        private int age;
        public User(int userId, String name, Date birth, int age) {
                this.userId = userId;
                this.name = name;
                this.birth = birth;
                this.age = age;
        }
        /**
         * 显示用户基本信息
         */
        @Override
        public String toString() {
            return "用户信息 [userId=" + userId + ", name=" + name + ",
                birth=" + birth + ", age=" + age + "]";
        }
}
```

（2）beans.xml

```
<?xml version="1.0" encoding="UTF-8"?>
<beans xmlns="http://www.springframework.org/schema/beans"
    xmlns:xsi="http://www.w3.org/2001/XMLSchema-instance"
    xsi:schemaLocation="http://www.springframework.org/schema/beans
        http://www.springframework.org/schema/beans/spring-beans.xsd">
    <bean id="user" class="com.inspur.entity.User">
      <constructor-arg name="userId" value="11"></constructor-arg>
      <constructor-arg name="name" value="tom"></constructor-arg>
      <constructor-arg name="birth" ref="birthDate"></constructor-arg>
      <constructor-arg name="age" value="20"></constructor-arg>
    </bean>
    <bean id="birthDate" class="java.util.Date"></bean>
</beans>
```

（3）Test.java

```
package com.inspur.test;
```

```
import org.springframework.context.ApplicationContext;
import org.springframework.context.support.ClassPathXmlApplicationContext;
import com.inspur.entity.User;
public class Test {
    public static void main(String[] args) {
        ApplicationContext applicationContext=new
        ClassPathXmlApplicationContext("beans.xml");
        User user=(User)applicationContext.getBean("user");
        System.out.println(user.toString());
    }
}
```

选中 Test.java 文件，单击鼠标右键，选择“Run As”菜单中的子菜单“1 Java Application”，运行程序后，在控制台上显示的结果如图 3-1 所示。

```
用户信息 [userId=11, name=tom, birth=Fri Sep 20 16:51:06 CST 2019, age=20]
```

图 3-1　验证构造器依赖注入案例运行结果

【实例 3.2】验证 setter 依赖注入。

实例 3.2

具体源代码如下所示（代码详见 Spring5_Ch03_inject02\src\com\inspur\entity\User.java；Spring5_Ch03_inject02\src\beans.xml；Spring5_Ch03_inject02\src\com\inspur\test\Test.java）。

（1）User.java

```
package com.inspur.entity;
import java.util.Date;
public class User {
    private int userId;
    private String name;
    private Date birth;
    private int age;
    public int getUserId() {
        return userId;
    }
    public void setUserId(int userId) {
        this.userId = userId;
    }
    public String getName() {
        return name;
    }
    public void setName(String name) {
        this.name = name;
    }
    public Date getBirth() {
        return birth;
    }
    public void setBirth(Date birth) {
        this.birth = birth;
    }
    public int getAge() {
        return age;
    }
    public void setAge(int age) {
```

```
            this.age = age;
        }
        /**
         * 显示用户基本信息
         */
        @Override
        public String toString() {
            return "用户信息 [userId=" + userId + ", name=" + name + ",
                birth=" + birth + ", age=" + age + "]";
        }
    }
```

（2）beans.xml

```
<?xml version="1.0" encoding="UTF-8"?>
<beans xmlns="http://www.springframework.org/schema/beans"
    xmlns:xsi="http://www.w3.org/2001/XMLSchema-instance"
    xsi:schemaLocation="http://www.springframework.org/schema/beans
    http://www.springframework.org/schema/beans/spring-beans.xsd">
    <bean id="user" class="com.inspur.entity.User">
         <property name="userId" value="11"></property>
         <property name="name" value="tom"></property>
         <property name="age" value="20"></property>
         <property name="birth" ref="birthDate"></property>
    </bean>
    <bean id="birthDate" class="java.util.Date"></bean>
</beans>
```

（3）Test.java

```
package com.inspur.test;
import org.springframework.context.ApplicationContext;
import org.springframework.context.support.ClassPathXmlApplicationContext;
import com.inspur.entity.User;
public class Test {
    public static void main(String[] args) {
        ApplicationContext applicationContext=new
ClassPathXmlApplicationContext("beans.xml");
        User user=(User)applicationContext.getBean("user");
        System.out.println(user.toString());
    }
}
```

选中 Test.java 文件，单击鼠标右键，选择“Run As”菜单中的子菜单“1 Java Application”，运行程序后，在控制台上显示的结果如图 3-2 所示。

```
用户信息 [userId=11, name=tom, birth=Fri Sep 20 16:51:06 CST 2019, age=20]
```

图 3-2　验证 setter 依赖注入案例运行结果

关于两种依赖注入类型的对比总结如下。

这两种注入方式都是非常有用的，没有绝对的好坏，只是适应的场景有所不同。相比之下，基于 setter 的注入比较常用，具有如下优点：基于 setter 的注入的方式与传统的 JavaBean 的写法更相似，程序员更容易理解、接受。且对于 JavaBean 类中的复杂的依赖关系，如果采用构造注入，会导致构造器过于臃肿，难以阅读。

p 命名空间注入

3.2.4　p 命名空间注入

我们可以使用 p 命名空间进行更简洁的 XML 配置。越来越多的 XML 文件采用属性而非子元素配置信息。Spring 从 2.5 版本开始引入了一个新的 p 命名空间，可以通过元素属性的方式配置 Bean 的属性，使用 p 命名空间后，基于 XML 的配置方式将进一步简化。

可以通过在头文件中加入 xmlns:p="http://www.springframework.org/schema/p"来实现。

例如：

```
<?xml version="1.0" encoding="UTF-8"?>
<beans xmlns="http://www.springframework.org/schema/beans"
   xmlns:p="http://www.springframework.org/schema/p"
   xmlns:xsi="http://www.w3.org/2001/XMLSchema-instance"
   xsi:schemaLocation="http://www.springframework.org/schema/beans
   http://www.springframework.org/schema/beans/spring-beans.xsd">
```

在实例 3.2 验证 setter 依赖注入实例中，User.java 在 Spring 配置文件中配置 bean 如下。

```
<?xml version="1.0" encoding="UTF-8"?>
<beans xmlns="http://www.springframework.org/schema/beans"
   xmlns:p="http://www.springframework.org/schema/p"
   xmlns:xsi="http://www.w3.org/2001/XMLSchema-instance"
   xsi:schemaLocation="http://www.springframework.org/schema/beans
   http://www.springframework.org/schema/beans/spring-beans.xsd">
      <bean id="user" class="com.inspur.entity.User" p:userId="11" p:name="tom"
          p:age="20" p:birth-ref="birthDate">
      </bean>
      <bean id="birthDate" class="java.util.Date"></bean>
</beans>
```

选中该实例 3.2 中的 Test.java 文件，单击鼠标右键，选择“Run As”菜单中的子菜单“1 Java Application”，运行程序后，在控制台上显示的结果如图 3-2 所示。

【程序解析】使用 p 的命名空间注入，其本质上还是使用 setter 方法进行注入依赖，使用 p 命名空间注入会使 XML 配置更简洁。

3.3　配置依赖

配置依赖

下面介绍依赖的配置。

1．简单属性的依赖的配置

如果 JavaBean 的简单属性为 Java 的基本数据类型 byte、short、int、long、float、double、boolean、char 或 String 时，在 Spring 配置文件中配置可以使用<value/>，用于指定字符串、基本类型的属性值。Spring 使用 XML 解析器解析这些数据，然后利用 java.beans.PropertyEditor 完成类型转换，从 String 类型转换到所需要的参数值类型。

2．配置合作者 Bean 的依赖的配置

如果 JavaBean 的属性值是容器中的另一个 Bean 实例时，在 Spring 配置文件中配置 Bean

时，可以使用 ref 属性指定另一个 bean 实例的 id。语法为<bean name = " "ref = " ">。

3. 集合属性的依赖的配置

如果 JavaBean 的属性值的类型是 Collection 的 List、Set、Map 和 Properties 集合时，则在 Spring 配置文件中可以使用集合元素，<list/><set/><map/>和<props/>可用来设置类型为 List、Set、Map 和 Properties 的集合属性值。

4. 注入嵌套 bean

如果某个 JavaBean 所依赖的 JavaBean 对象不想被 Spring 容器直接访问，可以使用嵌套 bean。在 Spring 配置文件中<bean/>用来定义嵌套 bean，嵌套 bean 只对嵌套它的外部 bean 有效，Spring 容器无法直接访问嵌套 bean，因此定义嵌套 bean 时无须指定 id 属性。

配置依赖案例-1

配置依赖案例-2

【实例 3.3】验证配置依赖案例。为了更好地说明如何配置依赖，读者可扫描二维码查看补充案例的视频。该案例在 Spring 5 及以下版本均能运行。

具体源代码如下所示（代码详见 Spring5_Ch03_inject03\src\com\inspur\Address.java；Spring5_Ch03_inject03\src\com\inspur\User.java；Spring5_Ch03_inject03\src\config\beans.xml；Spring5_Ch03_inject03\src\com\inspur\test\Test.java）。

（1）Address.java

```
package com.inspur;
/**
 *地址
 *
 */
public class Address {
    /**
     * 省份
     */
    private String prov;
    /**
     * 城市
     */
    private String city;
    /**
     * 街道
     */
    private String street;
    public void setProv(String prov) {
        this.prov = prov;
    }
    public void setCity(String city) {
        this.city = city;
    }
    public void setStreet(String street) {
        this.street = street;
    }
```

```
    @Override
    public String toString() {
        return "Address [prov=" + prov + ", city=" + city + ", street=" + street + "]";
    }
}
```

（2）User.java

```
package com.inspur;
import java.util.Arrays;
import java.util.List;
import java.util.Map;
import java.util.Properties;
import java.util.Set;
/**
 * 用户类
 */
public class User {
    /**
     * 用户编号
     */
    private int id;
    /**
     * 姓名
     */
    private String name;
    /**
     * 年龄
     */
    private int age;
    /**
     * 工资
     */
    private double sal;
    /**
     * 数组
     */
    private String stunoArray[];
    /**
     * List
     */
    private List lists;
    /**
     * Set
     */
    private Set sets;
    /**
     * Map
     */
    private Map scoreMap;
    /**
     * Properties 集合
     */
    private Properties props;
```

```
    /**
     * 地址
     */
    private Address address;
    /**
     * setter 方法
     * @param id
     */
    public void setId(int id) {
        this.id = id;
    }
    public void setName(String name) {
        this.name = name;
    }
    public void setAge(int age) {
        this.age = age;
    }
    public void setSal(double sal) {
        this.sal = sal;
    }
    public void setStunoArray(String[] stunoArray) {
        this.stunoArray = stunoArray;
    }
    public void setLists(List lists) {
        this.lists = lists;
    }
    public void setSets(Set sets) {
        this.sets = sets;
    }
    public void setScoreMap(Map scoreMap) {
        this.scoreMap = scoreMap;
    }
    public void setProps(Properties props) {
        this.props = props;
    }
    public void setAddress(Address address) {
        this.address = address;
    }
    @Override
    public String toString() {
        return "User [id=" + id + ", name=" + name + ", age=" + age + ",
        sal="+ sal + ", stunoArray=" + Arrays.toString(stunoArray) + ",
        lists=" + lists + ", sets=" + sets + ", scoreMap=" + scoreMap + ",
        props=" + props + ", address=" + address + "]";
    }
}
```

（3）beans.xml

```
<?xml version="1.0" encoding="UTF-8"?>
<beans xmlns="http://www.springframework.org/schema/beans"
    xmlns:xsi="http://www.w3.org/2001/XMLSchema-instance"
    xsi:schemaLocation="http://www.springframework.org/schema/beans
    http://www.springframework.org/schema/beans/spring-beans.xsd">
    <bean id="address" class="com.inspur.Address">
```

```
        <property name="prov" value="山东省"/>
        <property name="city" value="济南市"/>
        <property name="street" value="历下区"/>
    </bean>
    <bean id="user" class="com.inspur.User">
         <property name="id" value="1100"></property>
         <property name="name" value="scott"></property>
         <property name="age" value="30"></property>
         <property name="sal" value="9000.0"></property>
         <property name="stunoArray">
           <list>
               <value>J17001</value>
               <value>J17002</value>
               <value>J17003</value>
           </list>
         </property>
         <property name="lists">
           <list>
              <value>Java EE 课程</value>
              <value>大数据课程</value>
              <value>Java 课程</value>
           </list>
         </property>
         <property name="sets">
           <set>
              <value>阅读</value>
              <value>打游戏</value>
           </set>
         </property>
         <property name="scoreMap">
           <map>
               <entry key="english" value="90"/>
               <entry key="math" value="100"/>
            </map>
         </property>
         <property name="props">
             <props>
                <prop key="ironAxe">com.inspur.impl.IronAxe</prop>
                <prop key="stoneAxe">com.inspur.impl.StoneAxe</prop>
             </props>
         </property>
         <property name="address">
             <!-- 内部 bean：该 bean 只是在 User 中用，其他类访问不到 -->
             <bean class="com.inspur.Address">
                <property name="prov" value="山东省"/>
                <property name="city" value="济南市"/>
                <property name="street" value="历下区"/>
             </bean>
         </property>
      </bean>
  </beans>
```

（4）Test.java

```
package com.inspur.test;
import org.springframework.context.ApplicationContext;
import org.springframework.context.support.ClassPathXmlApplicationContext;
import com.inspur.User;
public class Test {
      public static void main(String[] args) {
      ApplicationContext  ac=new
       ClassPathXmlApplicationContext("config/beans.xml");
         User user=(User)ac.getBean("user");
             System.out.println(user);
        }
}
```

选中 Test.java 文件，单击鼠标右键，选择“Run As”菜单中的子菜单“1 Java Application”，运行程序后，在控制台上显示的结果如图 3-3 所示。

```
User [id=1100, name=scott, age=30, sal=9000.0,stunoArray=[J17001, J17002, J17003],
lists=[Java EE课程,大数据课程, Java课程], sets=[阅读, 打游戏],scoreMap={english=90, math=100},
props={ironAxe=com.inspur.impl.IronAxe, stoneAxe=com.inspur.impl.StoneAxe},
address=Address [prov=山东省, city=济南市, street=历下区]]
```

图 3-3　验证配置依赖案例运行结果

【程序解析】在 beans.xml 中 address 属性设置的值为内部 bean，该内部 bean 只在该 User 类的对象中起作用。如果 address 属性设置的值引用了外部 bean，则可修改为如下代码。

```
<property name="address" ref="address">
```

3.4 Bean 的作用域

Bean 的作用域

Spring 容器可以创建 JavaBean 实例，也可以指定其特定的作用域。在 Spring 配置文件中可以使用 Scope 来声明对象的作用域或者存活时间，即 Spring 容器在对象进入其相应的 Scope 之前，生成并配置这些对象，在该对象不再处于这些 Scope 的限定之后，Spring 容器通常会销毁这些对象。下面对 Bean 的作用域进行讲解。

3.4.1 作用域的种类

在 Spring 5 中 Spring Framework 支持 6 种 Bean 的作用域，其中只有在使用 Web 感知时才可用 ApplicationContext。这 6 种作用域及其说明如表 3-2 所示。

表 3-2　Bean 的作用域

范围	描述
singleton	单实例作用域，这是 Spring 容器默认的作用域，使用 singleton 作用域生成的是单实例，在整个 Spring 容器中仅保留一个实例对象供所有调用者共享引用。单例模式对于那些无会话状态的 Bean（如辅助工具类、DAO 组件、业务逻辑组件等）是最理想的选择

续表

范围	描述
prototype	原型模式，这是多实例作用域，针对每次不同的请求，Bean 容器均会生成一个全新的 Bean 实例以供调用者使用。prototype 作用域适用于那些需要保持会话状态的 Bean 实例。特别值得注意的是，Spring 不能对 prototype Bean 的整个生命周期负责，容器在初始化、装配好一个 prototype 实例后，将它交给客户端，随后就对该 prototype 实例不闻不问了。因此，客户端要负责 prototype 实例的生命周期管理
request	针对每次 HTTP 请求，Spring 容器都会根据 Bean 的定义创建一个全新的 Bean 实例，且该 Bean 实例仅在当前 HTTP request 内有效，因此我们可以根据需要放心地更改所建实例的内部状态，而其他请求中根据 Bean 定义创建的实例，将不会看到这些特定于某个请求的状态变化。当处理请求结束，request 作用域的 Bean 实例将被销毁。该作用域仅在基于 Web 的 Spring Application Context 情形下有效
session	针对某个 HTTP Session，Spring 容器会根据 Bean 定义创建一个全新的 Bean 实例，且该 Bean 实例仅在当前 HTTP Session 内有效。与 request 作用域一样，我们可以根据需要放心地更改所创建实例的内部状态，而别的 HTTP Session 中根据 Bean 定义创建的实例，将不会看到这些特定于某个 HTTP Session 的状态变化。当 HTTP Session 最终被废弃的时候，在该 HTTP Session 作用域内的 Bean 实例也会被废弃。该作用域仅在基于 Web 的 Spring ApplicationContext 情形下有效
application	将单个 Bean 定义的作用域限定为 ServletContext。为每个 ServletContext 对象创建一个实例。该作用域仅在基于 Web 的 Spring ApplicationContext 情形下有效
websocket	将单个 Bean 定义的作用域限定为 WebSocket。为每个 WebSocket 对象创建一个实例。该作用域仅在基于 Web 的 Spring ApplicationContext 情形下有效

在上述 6 种作用域中，singleton 和 prototype 是最常用的两种，下面针对这两种作用域做详细讲解。

3.4.2　singleton 作用域

singleton 是 Spring 容器中的默认 Bean 范围。它告诉 Spring 容器，为整个 Web 应用程序只创建和管理一个 Bean 类的实例。此单个实例存储在此类单例 Bean 的缓存中，并且该 Bean 的所有后续请求和引用都将返回缓存的实例。图 3-4 显示了 singleton 作用域的工作方式。

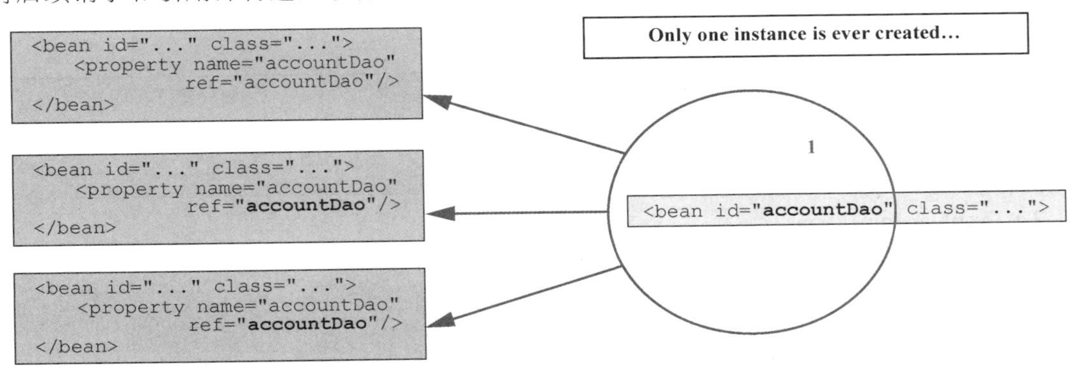

图 3-4　singleton 作用域的工作方式

要将 Bean 定义为 XML 中的单例，示例代码如下。

```
<bean id="accountService" class="com.something.DefaultAccountService"/>

<bean id="accountService" class="com.something.DefaultAccountService"
scope="singleton"/>
```

以上两段代码效果相同，Bean 的作用域中可以省略 scope="singleton"。

3.4.3　prototype 作用域

prototype 作用域是指应用程序代码每次对 Bean 发出请求时，Spring 容器都会创建一个新的 Bean 实例来处理请求。图 3-5 显示了 prototype 作用域的工作方式。

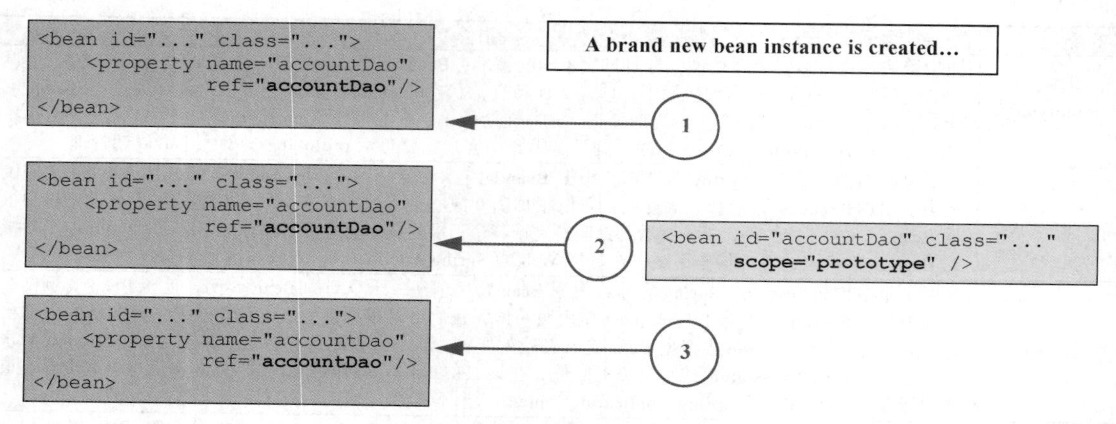

图 3-5　prototype 作用域的工作方式

以下示例将 Bean 定义为 XML 原型。

```
<bean id="accountService" class="com.something.DefaultAccountService"
scope="prototype"/>
```

【实例 3.4】验证 Bean 的作用域案例。

具体源代码如下所示（代码详见 Spring5_Ch03_scope\src\com\inspur\dao\EmpDao.java；Spring5_Ch03_scope\src\com\inspur\dao\impl\EmpDaoImpl.java；Spring5_Ch03_scope\src\com\inspur\service\EmpService.java；Spring5_Ch03_scope\src\com\inspur\service\impl\EmpServiceImpl.java；Spring5_Ch03_scope\src\beans.xml；Spring5_Ch03_scope\src\com\inspur\test\Test.java）。

（1）EmpDao.java

```
package com.inspur.dao;
public interface EmpDao {
    public int saveEmp();
}
```

（2）EmpDaoImpl.java

```
package com.inspur.dao.impl;
import com.inspur.dao.EmpDao;
public class EmpDaoImpl implements EmpDao{
    public int saveEmp() {
        System.out.println("保存新员工成功! ");
        return 0;
    }
}
```

（3）EmpService.java

```
package com.inspur.service;
public interface EmpService {
    public int saveEmp();
}
```

（4）EmpServiceImpl.java

```
package com.inspur.service.impl;
```

```
import com.inspur.dao.EmpDao;
import com.inspur.service.EmpService;
public class EmpServiceImpl implements EmpService{
private EmpDao empDao;
public EmpServiceImpl() {
    System.out.println("EmpServiceImpl 对象创建了");
}
public void setEmpDao(EmpDao empDao) {
    this.empDao = empDao;
}
public int saveEmp() {
    return empDao.saveEmp();
}
}
```

（5）beans.xml

```
<?xml version="1.0" encoding="UTF-8"?>
<beans xmlns="http://www.springframework.org/schema/beans"
  xmlns:xsi="http://www.w3.org/2001/XMLSchema-instance"
  xsi:schemaLocation="http://www.springframework.org/schema/beans
   http://www.springframework.org/schema/beans/spring-beans.xsd">
  <bean id="empDao" class="com.inspur.dao.impl.EmpDaoImpl"></bean>
  <bean id="empService" class="com.inspur.service.impl.EmpServiceImpl"
scope="singleton">
      <property name="empDao" ref="empDao"></property>
 </bean>
</beans>
```

（6）Test.java

```
package com.inspur.test;
import org.springframework.context.ApplicationContext;
import org.springframework.context.support.ClassPathXmlApplicationContext;
import com.inspur.service.EmpService;
public class Test {
public static void main(String[] args) {
    ApplicationContext ac=new ClassPathXmlApplicationContext("beans.xml");
    EmpService empService1=(EmpService)ac.getBean("empService");
    EmpService empService2=(EmpService)ac.getBean("empService");
    if(empService1==empService2) {
        System.out.println("同一个实例");
    }else {
        System.out.println("不是同一个实例");
    }
 }
 }
```

选中 Test.java 文件，单击鼠标右键，选择“Run As”菜单中的子菜单“1 Java Application”，运行程序后，在控制台上显示的结果如图 3-6 所示。

```
EmpServiceImpl对象创建了
同一个实例
```

图 3-6　验证 Bean 的作用域为 singleton 的案例运行结果

【程序解析】在 beans.xml 中 EmpServiceImpl 对象的 scope 范围配置为 singleton，通过图 3-6 可以看出应用程序对该 Bean 发出了两次请求，但 Spring 容器只是返回一个实例。

如果 beans.xml 中 EmpServiceImpl 对象的 scope 范围配置为 prototype，代码如下所示。

```
<bean id="empService" class="com.inspur.service.impl.EmpServiceImpl"
scope="prototype">
    <property name="empDao" ref="empDao"></property>
</bean>
```

程序运行结果如图 3-7 所示。运行结果中该 Bean 对象针对两次请求，创建了两个对象。

```
EmpServiceImpl对象创建了
EmpServiceImpl对象创建了
不是同一个实例
```

图 3-7　验证 Bean 的作用域为 prototype 的案例运行结果

项目中 DAO 组件、业务逻辑组件等建议其作用域设置为 singleton。

3.5　Bean 的自动装配

Bean 的自动装配

下面讲解 Spring 容器的第 2 种 Bean 的装配模式——自动装配，也就是说 Spring 容器可以自动装配协作 Bean 之间的依赖关系，即在 Spring 配置文件中无须使用 ref 显式指定依赖 Bean。由 BeanFactory 检查 XML 配置文件内容，根据某种规则，为主调用者 Bean 注入依赖关系。

自动装配具有以下优点：自动装配可以大大减少指定属性或构造函数参数的需要，从而减少了配置文件的工作量。

随着 Bean 对象的扩展，自动装配可以更新配置。例如，如果我们需要向类中添加依赖项，则无须修改配置即可自动满足该依赖项。

自动装配虽然可以大大减少配置的工作量，但降低了依赖关系的透明性和清晰性。所以编码阶段建议使用自动装配提高开发效率，但维护阶段建议使用手动装配的形式体现 Bean 对象间的依赖关系，从而提高代码的可读性。

Spring 的自动装配功能具有 4 种模式。表 3-3 描述了 4 种自动装配模式。

表 3-3　　Bean 的自动装配模式

模式	说明
no	无自动装配（默认）。Bean 引用必须由 ref 元素定义。对于较大的部署，建议不要更改默认设置，明确指定协作者可以提供更好的控制和代码清晰度。在某种程度上，它也是系统架构的一种文档形式
byName	根据属性名自动装配。此选项将检查容器并根据名字查找与属性完全一致的 Bean，并将其与属性自动装配。例如，在 Bean 定义中将 autowire 设置为 byName，而该 Bean 包含 master 属性（同时提供 setMaster(..) 方法），Spring 就会查找名为 master 的 Bean 定义，并用它来装配 master 属性
byType	如果容器中存在一个与指定属性类型相同的 Bean，那么将与该属性自动装配。如果存在多个该类型的 Bean，那么将会抛出异常，并指出不能使用 byType 方式进行自动装配。若没有找到相匹配的 Bean，则什么事都不发生，属性也不会被设置
constructor	与 byType 类似，不同之处在于它应用于构造器参数。如果在容器中没有找到与构造器参数类型一致的 Bean，那么将会抛出异常

使用基于 XML 的配置时，Spring 的自动装配可通过<beans/>的 default-autowire 属性指定，也可通过<bean/>的 autowire 属性指定。

实际应用中，我们常使用 byName、byType 两种自动装配方式，下面来看实例 3.5。

【实例 3.5】 验证 Bean 的自动装配方式 byName 的案例。

具体源代码如下所示（代码详见 Spring5_Ch03_autowire\src\com\inspur\entity\Bean1.java；Spring5_Ch03_autowire\src\com\inspur\entity\Bean2.java；Spring5_Ch03_autowire\src\com\inspur\entity\Bean3.java；Spring5_Ch03_scope\src\config\beans.xml；Spring5_Ch03_scope\src\com\inspur\test\Test.java）。

（1）Bean1.java

```
package com.inspur.entity;
public class Bean1 {
    private String name;
    private int age;
    public void setName(String name) {
        this.name = name;
    }
    public void setAge(int age) {
        this.age = age;
    }
}
```

（2）Bean2.java

```
package com.inspur.entity;
public class Bean2 {
    private String email;
    public void setEmail(String email) {
        this.email = email;
    }
}
```

（3）Bean3.java

```
package com.inspur.entity;
public class Bean3 {
    private Bean1 bean1;
    private Bean2 bean2;
    public Bean1 getBean1() {
        return bean1;
    }
    public void setBean1(Bean1 bean1) {
        this.bean1 = bean1;
    }
    public Bean2 getBean2() {
        return bean2;
    }
    public void setBean2(Bean2 bean2) {
        this.bean2 = bean2;
    }
}
```

（4）beans.xml

```
<?xml version="1.0" encoding="UTF-8"?>
<beans xmlns="http://www.springframework.org/schema/beans"
    xmlns:xsi="http://www.w3.org/2001/XMLSchema-instance"
    xsi:schemaLocation="http://www.springframework.org/schema/beans
          http://www.springframework.org/schema/beans/spring-beans.xsd"
default-autowire="byName">
   <bean id="bean1" class="com.inspur.entity.Bean1"/>
   <bean id="bean2" class="com.inspur.entity.Bean2"/>
   <bean id="bean3" class="com.inspur.entity.Bean3" />
</beans>
```

（5）Test.java

```
package com.inspur.test;
import org.springframework.context.ApplicationContext;
import org.springframework.context.support.
ClassPathXmlApplicationContext;
import com.inspur.entity.Bean3;
public class Test {
     public static void main(String[] args) {
ApplicationContext ac=new ClassPathXmlApplicationContext("config/beans.xml");
          Bean3 bean3=(Bean3)ac.getBean("bean3");
          System.out.println(bean3.getBean1());
          System.out.println(bean3.getBean2());
     }
}
```

选中 Test.java 文件，单击鼠标右键，选择“Run As”菜单中的子菜单“1 Java Application”，运行程序后，在控制台上显示的结果如图 3-8 所示。

```
com.inspur.entity.Bean1@5f341870
com.inspur.entity.Bean2@553f17c
```

图 3-8 验证 Bean 的自动装配方式为 byName 案例运行结果

【案例解析】

该案例是按照属性名称进行自动装配的，如果将上述配置文件 beans.xml 修改为：

```
<bean id="bean1" class="com.inspur.entity.Bean1" autowire="byName"/>
```

运行 Test.java 文件后，在控制台上运行结果和图 3-8 相同。

【实例 3.6】验证 Bean 的自动装配方式 byType 的案例。

具体源代码如下所示（代码详见 Spring5_Ch03_autowire\src\com\inspur\entity\Bean1.java；Spring5_Ch03_autowire\src\com\inspur\entity\Bean2.java；Spring5_Ch03_autowire\src\com\inspur\entity\Bean3.java；Spring5_Ch03_scope\src\config\beans.xml；Spring5_Ch03_scope\src\com\inspur\test\Test.java）。

其中 Bean1.java、Bean2.java 参照实例 3.5 中的代码。

（1）Bean3.java 修改如下

```
package com.inspur.entity;
public class Bean3 {
```

```
    private Bean1 bean31;
    private Bean2 bean32;
    public Bean1 getBean31() {
        return bean31;
    }
    public void setBean31(Bean1 bean31) {
        this.bean31 = bean31;
    }
    public Bean2 getBean32() {
        return bean32;
    }
    public void setBean2(Bean2 bean32) {
        this.bean32 = bean32;
    }
}
```

（2）beans.xml 修改如下

```
<?xml version="1.0" encoding="UTF-8"?>
<beans xmlns="http://www.springframework.org/schema/beans"
    xmlns:xsi="http://www.w3.org/2001/XMLSchema-instance"
    xsi:schemaLocation="http://www.springframework.org/schema/beans
    http://www.springframework.org/schema/beans/spring-beans.xsd"
            default-autowire="byType">
    <bean id="bean1" class="com.inspur.entity.Bean1" />
    <bean id="bean2" class="com.inspur.entity.Bean2"/>
    <bean id="bean3" class="com.inspur.entity.Bean3" autowire="byType" />
</beans>
```

（3）Test.java 修改如下

```
package com.inspur.test;
import org.springframework.context.ApplicationContext;
import org.springframework.context.support.ClassPathXmlApplicationContext;
import com.inspur.entity.Bean3;
public class Test {
    public static void main(String[] args) {
        ApplicationContext ac=new
ClassPathXmlApplicationContext("config/beans.xml");
        Bean3 bean3=(Bean3)ac.getBean("bean3");
        System.out.println(bean3.getBean31());
        System.out.println(bean3.getBean32());
    }
}
```

选中 Test.java 文件，单击鼠标右键，选择“Run As”菜单中的子菜单“1 Java Application”，运行程序后，在控制台上显示的结果如图 3-9 所示。

```
com.inspur.entity.Bean1@73ad2d6
com.inspur.entity.Bean2@7085bdee
```

图 3-9　验证 Bean 的自动装配方式为 byType 案例运行结果

3.6　Bean 的基于 Annotation 的装配

在 Spring 中，可以使用基于 XML 配置的形式完成 Bean 的装配，如果应用中有很多的 Bean，则 XML 配置文件会比较臃肿，后期软件维护或升级会有一定的困难。为此 Spring 提供了第三种 Bean 的装配形式：基于注解（Annotation）的装配。这也是常用的装配方式。

Spring 中定义的关于 Bean 装配的常用注解，可以分为创建对象的注解和注入数据的注解两类。

3.6.1 用于创建对象的注解

用于创建对象的注解

@Component 可标注一个普通的 Spring Bean 类，把普通 Bean 实例化到 Spring 容器中，相当于 XML 配置文件中的<bean id="" class=""/>。该注解的 value 属性可用于指定 Bean 的 id。当不写时，默认值是当前类名首字母小写。

① @Controller：是@Component 的衍生注解，一般用于表现层，标注一个控制器组件类。

② @Service：是@Component 的衍生注解，一般用于业务逻辑层，标注一个业务逻辑组件类。

③ @Repository：是@Component 的衍生注解，一般用于持久层，标注一个 DAO 组件类。

@Controller、@Service、@Repository 三个衍生注解的作用及属性与@Component 相同。它们只是 Spring 框架提供了语义上的分层，所以注解之间可以互换，但在实际应用中还是使用语义上分层的注解比较好。

下面通过案例演示如何通过这些注解来装配 Bean。

【实例 3.7】验证@Component 的案例。

具体源代码如下所示（代码详见 Spring5_Ch03_annotationIOC\src\com\inspur\service\CustomerService.java；Spring5_Ch03_annotationIOC\src\com\inspur\service\impl\CustomerServiceImpl.java；Spring5_Ch03_annotationIOC\src\beans.xml；Spring5_Ch03_annotationIOC\src\com\inspur\test\Test.java）。

（1）CustomerService.java

```
package com.inspur.service;
public interface CustomerService {
   void saveCustomer();
}
```

（2）CustomerServiceImpl.java

```
package com.inspur.service.impl;
import org.springframework.stereotype.Component;
import com.inspur.service.CustomerService;
@Component(value="customerService")
public class CustomerServiceImpl implements CustomerService{
   @Override
   public void saveCustomer() {
   }
}
```

（3）beans.xml

```
<?xml version="1.0" encoding="UTF-8"?>
<beans xmlns="http://www.springframework.org/schema/beans"
    xmlns:xsi="http://www.w3.org/2001/XMLSchema-instance"
    xmlns:context="http://www.springframework.org/schema/context"
    xsi:schemaLocation="http://www.springframework.org/schema/beans
        http://www.springframework.org/schema/beans/spring-beans.xsd
        http://www.springframework.org/schema/context
    http://www.springframework.org/schema/context/spring-context.xsd">

<!--告知 Spring 容器自动扫描组件包，创建 beans 放到 Spring 容器中 -->
<context:component-scanbase-package="com.inspur"></context:component-scan>
</beans>
```

（4）Test.java

```
package com.inspur.test;
import org.springframework.beans.factory.BeanFactory;
import org.springframework.context.support.ClassPathXmlApplicationContext;
import com.inspur.dao.CustomerDao;
import com.inspur.service.CustomerService;
public class Test {
    public static void main(String[] args) {
        BeanFactory factory=new ClassPathXmlApplicationContext("beans.xml");
        CustomerService customerService=factory.getBean("customerService",
    CustomerService.class);
        System.out.println(customerService);
    }
}
```

选中 Test.java 文件，单击鼠标右键，选择“Run As”菜单中的子菜单“1 Java Application”，运行程序后，在控制台上显示的结果如图 3-10 所示。

```
com.inspur.service.impl.CustomerServiceImpl@4516af24
```

图 3-10　验证@Component 的案例运行结果

【案例解析】在 beans.xml 中配置<context:component-scanbase-package ="com.inspur"/>是告知 Spring 容器自动扫描组件包，识别 Bean 类中的注解，并创建 beans 放到 Spring 容器中。@Component 的作用是为了创建 Spring Bean 对象。该注解的 value 属性，可用于指定 Bean 的 id，当不写时，默认值是当前类名首字母小写。如注解@Component()和@Component(value= "customerServiceImpl")同效果。若@Component 只包括 value 一个属性，我们可以将 value 属性名省略，将之简写为@Component("customerServiceImpl")。该实例需要图 3-11 所示的 JAR 包支持。

```
commons-logging-1.2.jar
spring-aop-5.0.2.RELEASE.jar
spring-beans-5.0.2.RELEASE.jar
spring-context-5.0.2.RELEASE.jar
spring-core-5.0.2.RELEASE.jar
spring-expression-5.0.2.RELEASE.jar
```

图 3-11　实例 3.7 所需的 JAR 包

【实例 3.8】验证@Repository@Service 的案例。

实例 3.8

具体源代码如下所示。其中 beans.xml 参见实例 3.7。

（1）CustomerDao.java

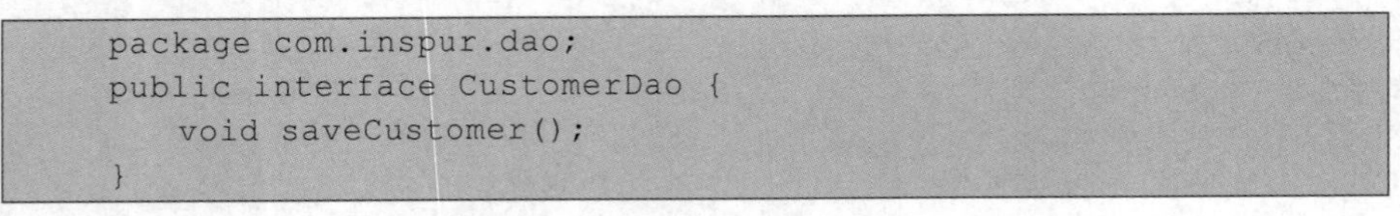

```
package com.inspur.dao;
public interface CustomerDao {
    void saveCustomer();
}
```

（2）CustomerDaoImpl.java

```
package com.inspur.dao.impl;
import org.springframework.stereotype.Repository;
import com.inspur.dao.CustomerDao;
@Repository("customerDao")
public class CustomerDaoImpl implements CustomerDao{
    public void saveCustomer() {
        System.out.println("保存客户信息");
    }
}
```

（3）CustomerService.java

参见实例 3.7。

（4）CustomerServiceImpl.java

修改如下。

```
package com.inspur.service.impl;
import org.springframework.stereotype.Service;
import com.inspur.dao.CustomerDao;
import com.inspur.service.CustomerService;
@Service(value="customerService")
public class CustomerServiceImpl implements CustomerService{
    private CustomerDao customerDao;
    @Override
    public void saveCustomer() {
        customerDao.saveCustomer();
    }
}
```

（5）Test.java

```
package com.inspur.test;
import org.springframework.beans.factory.BeanFactory;
import org.springframework.context.support.ClassPathXmlApplication
Context;
import com.inspur.dao.CustomerDao;
import com.inspur.service.CustomerService;
public class Test {
    public static void main(String[] args) {
        BeanFactory factory=new ClassPathXmlApplicationContext("beans.xml");
        CustomerService customerService=factory.getBean("customerService",
    CustomerService.class);
        CustomerDao customerDao=factory.getBean("customerDao",
CustomerDao.class);
        System.out.println(customerService);
```

```
            System.out.println(customerDao);
        }
    }
```

选中 Test.java 文件，单击鼠标右键，选择“Run As”菜单中的子菜单“1 Java Application”，运行程序后，在控制台上显示的结果如图 3-12 所示。

```
com.inspur.service.impl.CustomerServiceImpl@8b87145
com.inspur.dao.impl.CustomerDaoImpl@6483f5ae
```

图 3-12 验证@Repository@Service 的案例运行结果

【案例解析】@Service 注解可以用 value 属性指定 Bean 的 id，如果不指定 Bean 的 id，则默认 Bean 的 id 为类名首字符小写，也就是 customerServiceImpl。如果@Service 只有一个属性，则 value 可以省略不写。

3.6.2 用于注入数据的注解

用于注入数据的注解

@AutoWired 用于注入数据的注解，数据类型为其他类型的 Bean。装配过程为，自动按照类型进行匹配，如果 Spring 容器中有唯一类型匹配成功，则可以直接注入。如果有多个同类型的匹配时，可用变量名称作为 Bean 的 id 从 Spring 容器中查找，如果找到也可以注入成功。如果没有匹配的，则报错。

下面通过案例演示如何通过这些注解来装配 Bean。

【实例 3.9】验证@AutoWired 的案例。

具体源代码如下所示（代码详见 Spring5_Ch03_annotationIOC-autowired\src\com\inspur\dao\CustomerDao.java；Spring5_Ch03_annotationIOC-autowired\src\com\inspur\dao\impl\CustomerDaoImpl.java；Spring5_Ch03_annotationIOC-autowired\src\com\inspur\service\CustomerService.java；Spring5_Ch03_annotationIOC-autowired\src\com\inspur\service\impl\CustomerServiceImpl.java；Spring5_Ch03_annotationIOC-autowired\src\beans.xml；Spring5_Ch03_annotationIOC-autowired\src\com\inspur\test\Test.java）。

其中 CustomerDao.java、CustomerDaoImpl.java、CustomerService.java、beans.xml 可参见实例 3.7～3.8。

（1）CustomerServiceImpl.java

```
package com.inspur.service.impl;
import org.springframework.beans.factory.annotation.Autowired;
import org.springframework.stereotype.Service;
import com.inspur.dao.CustomerDao;
import com.inspur.service.CustomerService;
@Service(value="customerService")
public class CustomerServiceImpl implements CustomerService{
    @Autowired
    private CustomerDao customerDao;
    public void saveCustomer() {
        customerDao.saveCustomer();
    }
```

```
    }
```

（2）Test.java

```
    package com.inspur.test;
    import org.springframework.beans.factory.BeanFactory;
    import org.springframework.context.support.ClassPathXmlApplicationContext;
    import com.inspur.service.CustomerService;
    public class Test {
        public static void main(String[] args) {
            BeanFactory factory=new ClassPathXmlApplicationContext("beans.xml");
            CustomerService customerService=factory.getBean("customerService",
        CustomerService.class);
            customerService.saveCustomer();
        }
    }
```

选中 Test.java 文件，单击鼠标右键，选择“Run As”菜单中的子菜单“1 Java Application”，运行程序后，在控制台上显示的结果如图 3-13 所示。

保存客户信息

图 3-13　验证@AutoWired 的案例运行结果

【程序解析】当使用@AutoWired 注解时，可以省略 Bean 对象中的 set 方法，如 CustomerServiceImpl 类中的 setCustomerDao 方法，就可以省略了。@AutoWired 注解中可以使用属性 required，该属性用于设置是否必须注入成功，取值为 true（默认值）/false。取值为 true 时，如果没有匹配的对象，则程序就会报错。

3.6.3　用于指定 Bean 作用域的注解

用于指定 Bean 作用域的注解

@Scope 可指定 Bean 的作用域范围，其范围取值和 XML 配置是一致的，默认值为 singleton。使用属性 value 可指定作用域范围。

下面通过案例演示如何使用 Scope 注解。

【实例 3.10】验证@Scope 的案例。

具体源代码如下所示（代码详见 Spring5_Ch03_01_annotationIOC-scope\src\com\inspur\dao\CustomerDao.java；Spring5_Ch03_01_annotationIOC-scope\src\com\inspur\dao\impl\CustomerDaoImpl.java；Spring5_Ch03_01_annotationIOC-scope\src\com\inspur\service\CustomerService.java；Spring5_Ch03_01_annotationIOC-scope\src\com\inspur\service\impl\CustomerServiceImpl.java；Spring5_Ch03_01_annotationIOC-scope\src\beans.xml；Spring5_Ch03_01_annotationIOC-scope\src\com\inspur\test\Test.java）。

其中 CustomerDao.java、CustomerDaoImpl.java、CustomerService.java、beans.xml 可参见实例 3.7～3.8。

（1）CustomerServiceImpl.java

修改如下：

```
    package com.inspur.service.impl;
    import org.springframework.beans.factory.annotation.Autowired;
```

```
import org.springframework.context.annotation.Scope;
import org.springframework.stereotype.Service;
import com.inspur.dao.CustomerDao;
import com.inspur.service.CustomerService;
@Service("customerService")
@Scope("prototype")
public class CustomerServiceImpl implements CustomerService{
    @Autowired
    private CustomerDao customerDao;
    @Override
    public void saveCustomer() {
        customerDao.saveCustomer();
    }
}
```

（2）Test.java

修改如下：

```
package com.inspur.test;
import org.springframework.context.ApplicationContext;
import org.springframework.context.support.ClassPathXmlApplicationContext;
import com.inspur.service.CustomerService;
public class Test {
    public static void main(String[] args) {
        ApplicationContext ac=new ClassPathXmlApplicationContext("beans.xml");
        CustomerService customerService01=(CustomerService)ac.getBean
            ("customerService");
        CustomerService customerService02=(CustomerService)ac.getBean
            ("customerService");
        if(customerService01==customerService02) {
            System.out.println("同一个实例");
        }else {
            System.out.println("不是同一个实例");
        }
    }
}
```

选中 Test.java 文件，单击鼠标右键，选择“Run As”菜单中的子菜单“1 Java Application”，运行程序后，在控制台上显示的结果如图 3-14 所示。

【程序解析】@Scope 注解可指定 Bean 的作用范围，默认值为 singleton。也就是说使用@Scope 和@Scope("singleton")效果相同。如果使用@Scope 替换 CustomerServiceImpl.java 文件中的@Scope("prototype")，则程序的运行结果如图 3-15 所示。

不是同一个实例

图 3-14　验证@Scope 的案例运行结果

同一个实例

图 3-15　验证@Scope 的案例运行结果

用于将外部的值动态注入 Bean

3.6.4　用于将外部的值动态注入 Bean

@Value 用于不通过 XML 配置文件注入属性的情况，通过@Value 可将外部的值动态注入 Bean 中，使用的情况有：注入普通字符串；注入操作系统属性；注入表达式结果；注入其他 Bean 属性；注入文件资源；注入 URL 资源。

在后面的实例中会用到@Resource 注解，这里补充说明一下@Resource 注解。

把@Resource 注解放在@Autowired 之后介绍，是因为它们的作用相似，都是自动装配 Bean 对象。

@Resource 的装配顺序如下。

① @Resource 后面没有任何内容，默认通过 name 属性去匹配 Bean，找不到再按 type 去匹配。

② 指定了 name 或者 type，则根据指定的类型去匹配 Bean。

③ 指定了 name 和 type，则根据指定的 name 和 type 去匹配 Bean，任何一处不匹配都将报错。

下面介绍@Autowired 和@Resource 注解的区别。

① @Autowired 默认按照 byType 方式进行 Bean 匹配，@Resource 默认按照 byName 方式进行 Bean 匹配。

② @Autowired 是 Spring 的注解，@Resource 是 Java EE 的注解，看导入注解时这两个注解的包名就一清二楚了。Spring 属于第三方，Java EE 则是 Java 自己的，因此建议使用@Resource 注解，以减少代码和 Spring 之间的耦合。

下面通过案例演示如何使用 Value 注解。

【实例 3.11】 验证@Value 的案例 1。

具体源代码如下所示（代码详见 Spring5_Ch03_annotationIOC-value01\src\com\inspur\dao\CustomerDao.java；Spring5_Ch03_annotationIOC-value01\src\com\inspur\dao\impl\CustomerDaoImpl.java；Spring5_Ch03_annotationIOC-value01\src\com\inspur\service\CustomerService.java；Spring5_Ch03_annotationIOC-value01\src\com\inspur\service\impl\CustomerServiceImpl.java；Spring5_Ch03_annotationIOC-value01\src\beans.xml；Spring5_Ch03_annotationIOC-value01\src\com\inspur\test\Test.java）。

其中 CustomerDao.java、CustomerDaoImpl.java、CustomerService.java、beans.xml 可参见实例 3.7～3.8，Test.java 可参见实例 3.9。

CustomerServiceImpl.java 修改如下。

```
package com.inspur.service.impl;
import javax.annotation.Resource;
import org.springframework.beans.factory.annotation.Value;
import org.springframework.stereotype.Service;
import com.inspur.dao.CustomerDao;
import com.inspur.service.CustomerService;
@Service(value="customerService")
public class CustomerServiceImpl implements CustomerService{
     @Resource(name="customerDao")
     private CustomerDao customerDao;
     //这是第一种方式：直接指定值
     @Value("oracle.jdbc.driver.OracleDriver")
     private String driver;
     @Value("jdbc:oracle:thin:@localhost:1521:orcl")
     private String url;
     @Value("scott")
```

```
        private String username;
        @Value("tiger")
        private String password;
        @Override
        public void saveCustomer() {
            System.out.println("数据库连接参数如下：");
            System.out.println(" driver="+driver+"\n url="+url+"\n username="+
        username+"\n password="+password);
            customerDao.saveCustomer();
        }
    }
```

选中 Test.java 文件，单击鼠标右键，选择“Run As”菜单中的子菜单“1 Java Application”，运行程序后，在控制台上显示的结果如图 3-16 所示。

```
数据库连接参数如下：
 driver=oracle.jdbc.driver.OracleDriver
 url=jdbc:oracle:thin:@localhost:1521:orcl
 username=scott
 password=tiger
保存客户信息
```

图 3-16　验证@Value 的案例 1 的运行结果

【程序解析】 该案例中使用@Value("oracle.jdbc.driver.OracleDriver")方式直接将指定的值注入，缺点为数据库连接的参数值是写在 Java 文件中的，不太好维护。

下面通过案例 3.12 将数据库连接的参数值写到 properties 文件中，以增强代码的维护性。

【案例 3.12】 验证@Value 的案例 2。

具体源代码如下所示（代码详见 Spring5_Ch03_annotationIOC-value02\src\com\inspur\dao\CustomerDao.java；Spring5_Ch03_annotationIOC-value02\src\com\inspur\dao\impl\CustomerDaoImpl.java；Spring5_Ch03_annotationIOC-value02\src\com\inspur\service\CustomerService.java；Spring5_Ch03_annotationIOC-value02\src\com\inspur\service\impl\CustomerServiceImpl.java；Spring5_Ch03_annotationIOC-value02\src\ConfigDB.properties；Spring5_Ch03_annotationIOC-value02\src\beans.xml；Spring5_Ch03_annotationIOC-value02\src\com\inspur\test\Test.java）。

其中 CustomerDao.java、CustomerDaoImpl.java、CustomerService.java 可参见实例 3.8，Test.java 可参见实例 3.9。

（1）CustomerServiceImpl.java

修改如下：

```
package com.inspur.service.impl;
import javax.annotation.Resource;
import org.springframework.beans.factory.annotation.Value;
import org.springframework.stereotype.Service;
import com.inspur.dao.CustomerDao;
import com.inspur.service.CustomerService;
@Service(value="customerService")
public class CustomerServiceImpl implements CustomerService{

    @Resource(name="customerDao")
    private CustomerDao customerDao;
    //下面是第二种方式：使用 EL 表达式
```

```
        @Value("${driver}")
        private String driver;
        @Value("${url}")
        private String url;
        @Value("${username}")
        private String username;
        @Value("${password}")
        private String password;
        @Override
        public void saveCustomer() {
            System.out.println("数据库连接参数如下：");
            System.out.println(" driver="+driver+"\n url="+url+"\n username="+
        username+"\n password="+password);
            customerDao.saveCustomer();
        }
    }
```

（2）ConfigDB.properties

```
driver=oracle.jdbc.driver.OracleDriver
url=jdbc:oracle:thin:@localhost:1521:orcl
username=scott
password=tiger
```

（3）beans.xml

修改如下：

```
<?xml version="1.0" encoding="UTF-8"?>
<beans xmlns="http://www.springframework.org/schema/beans"
    xmlns:xsi="http://www.w3.org/2001/XMLSchema-instance"
    xmlns:context="http://www.springframework.org/schema/context"
    xsi:schemaLocation="http://www.springframework.org/schema/beans
    http://www.springframework.org/schema/beans/spring-beans.xsd
    http://www.springframework.org/schema/context
    http://www.springframework.org/schema/context/spring-context.xsd">
    <context:component-scan base-package="com.inspur"></context:component-scan>
    <!-- 告知 Spring 加载 properties 文件的位置 -->
    <context:property-placeholder location="ConfigDB.properties"/>
</beans>
```

选中该 Test.java 文件，单击鼠标右键，选择“Run As”菜单中的子菜单“1 Java Application”，运行程序后，在控制台上显示的结果如图 3-16 所示。

【程序解析】该案例使用注入表达式结果的方式进行值的注入，如代码@Value("${driver}")，${driver}为 EL 表达式。该表达式的值需要在资源文件中得到解析，所以需要在 beans.xml 中配置<context:property-placeholder location="ConfigDB.properties"/>。作用为：告知 Spring 容器加载 properties 文件的位置，然后加载该文件，从而使 EL 表达式的值得到解析。

本章小结

本章首先讲解了依赖注入的概念及依赖注入的类型，读者需要掌握构造函数的依赖注入、基于 setter 的依赖注入和 p 命名空间注入。然后本章讲解了怎样配置依赖、Bean 的作用域和 Bean 的自动装配方式。最后本章还讲解了基于注解的装配，重点要掌握基于注解的装配。

习题

1. 简述依赖注入的类型。
2. 自动装配注入合作者 Bean，常用哪几种方式？
3. 简述 Bean 的作用域范围。

上机指导

1. 开发一个 Web 工程，并引入 Spring 框架，完成人养了狗和猫。

分析：Person 类{
　　　　　　养狗；
　　　　　　养猫；
　　　　　　}

测试：人养狗、养猫。

以上请使用基于 XML 配置形式完成。

2. 开发一个 Java 工程，并引入 Spring 框架，完成模拟用户登录功能。

以上请使用基于注解的装配形式完成。

04

第4章　面向切面编程

学习目标

- 理解 AOP 的概念
- 掌握 AOP 的相关术语
- 掌握 Spring 支持 AOP 的实现方式
- 掌握基于 XML 的配置方式实现 AOP 编程的方法
- 掌握基于注解的方式实现 AOP 编程的方法

AOP 概述和相关术语

4.1　AOP 概述

面向切面编程（Aspect Oriented Programming，AOP）可以说是面向对象编程（Object Oriented Programming，OOP）的补充和完善。OOP引入封装、继承和多态性等概念来建立一种对象层次结构，用以模拟公共行为的一个集合。当我们需要为分散的对象引入公共行为的时候，OOP 则显得无能为力。也就是说，OOP 允许用户定义从上到下的关系，但并不适合定义从左到右的关系，例如日志打印功能。日志代码往往水平地散布在所有对象层次中，而与它所散布到的对象的核心业务功能毫无关系。这种散布在各处的无关代码被称为横切（Cross-Cutting）代码，在 OOP 设计中，它导致了大量代码的重复，而且不利于各个模块的重用。

而 AOP 技术则恰恰相反，它利用一种称为“横切”的技术，剖开封装的对象内部，将那些影响了多个类的公共行为封装到一个可重用模块，并将其命名为“Aspect”，也就是“切面”。AOP 代表的是一个横向的关系，如果说“对象”是一个空心的圆柱体，其中封装的是对象的属性和行为，那么 AOP 的方法就仿佛一把利刃，将这些空心圆柱体剖开，以获得其内部的消息。然后它又以巧夺天工的“妙手”将这些剖开的切面复原，不留痕迹。

使用“横切”技术，AOP 把软件系统分为两个部分：核心关注点和横切关注点。业务处理的主要流程是核心关注点，与之关系不大的部分是横切关注点。横切关注点的一个特点是，它们经常发生在核心关注点的多处，而各处都基本相似，例如权限认证、日志、事务处理等。AOP 的作用在于分离系统中的各个关注点，将核心关注点和横切关注点分离开来。正如有学者所说，AOP 的核心思想就是“将应用程序中的商业逻辑同对其提供支持的通用服务进行分离”。

使用 AOP 的目的是能够将那些与业务无关，却为业务模块所共同调用的逻辑或责任（如事务处理、日志管理、权限认证等）封装起来，以便于减少系统的重复代码，降低模块间的耦合度，并有利于未来的可拓展性和可维护性。

AOP 的主要功能是：日志记录、性能统计、安全控制、事务处理、异常处理等。

Spring 的关键组件之一就是 AOP 框架。Spring IoC 容器不依赖于 AOP，这意味着我们也可以不使用 AOP，但 AOP 是对 Spring IoC 的补充，可以提供功能强大的中间件解决方案。

4.2　AOP 的相关术语

下面介绍主要的 AOP 术语。AOP 的相关术语包括切面、连接点、通知、切入点、AOP 代理、织入、目标对象等，下面分别进行介绍。

（1）切面（Aspect）：一个关注点的模块化，这个关注点实现可能横切多个对象。如关注点日志打印可以在 userManager 类中插入，也可以在 CustomerManager 类中插入。切面一般定义为一个 Java 类，事务管理是企业 Java 应用程序中横切关注的一个很好的例子。

（2）连接点（Joinpoint）：在程序执行过程中的一点，如方法的执行或异常的处理。在 Spring AOP 中，连接点始终代表方法的执行。

（3）通知（Advice）：在特定的连接点（Joinpoint），AOP 框架执行的动作。通知有多种类型，有关的通知类型将在后面具体讲解。

（4）切入点（Pointcut）：一个匹配连接点的正则表达式。当一个连接点匹配到切点时，一个关联到这个切点的特定的通知（Advice）会被执行。

（5）AOP 代理（AOP Proxy）：AOP 框架创建的对象，包含通知。在 Spring 中，AOP 代理可以是 JDK 动态代理或者 CGLIB 代理。

（6）织入（Weaving）：负责将切面和目标对象链接，以创建通知对象。这可以在编译时完成（如使用 AspectJ 编译器），也可以在运行时完成。Spring 和其他纯 Java AOP 框架一样，可以在运行时完成织入。

（7）目标对象（Target Object）：包含连接点的对象，也称作被通知或被代理对象。

4.3　AOP 代理

AOP代理和切入点的定义

Spring 使用两种机制实现 AOP 代理：一是使用 Java 的动态代理，即 java.lang.reflect.Proxy 类创建 AOP 代理；二是使用 CGLIB 库自动生成目标对象的子类，同时织入通知。

Spring AOP 默认将标准 JDK 动态代理用于 AOP 代理。该动态代理要求目标对象必须实现接口（只有这样才能创建代理），而 CGLIB 则没有这种限制。

Spring AOP 也可以使用 CGLIB 代理。这对于代理类接口不是必需的。如果业务对象未实现接口，则默认情况下使用 CGLIB。因为对接口而不是类进行编程是一种好习惯。业务类通常会实现一个或多个业务接口。在那些需要建立在接口上未声明的方法或需要将代理对象作为具体类型传递给方法的情况下（极少数情况），可以强制使用 CGLIB。

Spring 中的 AOP 代理由 Spring 的 IoC 容器负责生成、管理，其依赖关系也由 IoC 容器负责管理。

面向 AOP 编程只需要程序员定义以下 3 个部分。

（1）定义普通业务组件。

（2）定义切入点。

（3）定义通知（也叫增强处理）。

进行 AOP 编程的关键就在于定义切入点和通知。一旦定义了合适的切入点和通知，AOP 框架就会自动生成 AOP 代理。

4.4 AOP 实战

4.4.1 AspectJ 简介

在进行 AOP 前，需要先了解一下 AspectJ。

AspectJ 实际上是对 AOP 思想的一个实践，是一个基于 Java 语言的 AOP 框架，它提供了强大的 AOP 功能。自 Spring 2.0 以后，Spring AOP 就引入了对 AspectJ 的支持。在新版本的 Spring 框架中，也建议直接使用 AspectJ 进行编程。

使用 AspectJ 实现 AOP 编程有两种方式：第一种是基于 XML 配置文件的方式，为使用 Spring 配置文件来定义切入点和通知；第二种是基于注解（Annotation）的方式，为使用@Aspect、@Pointcut 等注解来标注切入点和通知。

4.4.2 Spring 通知的类型

在进行 AOP 实战之前，还需要了解 Spring 的通知类型。Spring 中的通知按照在目标类方法的连接点位置，可以分为以下 5 种类型。

① 前置通知（Before Advice）：在连接点前面执行，前置通知不会影响连接点的执行，除非此处抛出异常。

② 正常返回通知（After Returning Advice）：在连接点正常执行完后执行，如果连接点抛出异常，则不会执行。

③ 异常返回通知（After Throwing Advice）：在连接点抛出异常后执行。

④ 最终通知（After/Finally Advice）：在连接点执行完后执行，不管是正常执行完成还是抛出异常，都会执行返回通知中的内容。

⑤ 环绕通知（Around Advice）：环绕通知围绕在连接点前后，如一个方法调用的前后。这是最强大的通知类型，能在方法调用前后自定义一些操作。环绕通知还需要负责决定是继续处理 Joinpoint（调用 ProceedingJoinPoint 的 proceed 方法），还是中断执行。

4.4.3　切入点的定义

在 Spring AOP 中切入点（Pointcut）表达式大量使用 execution 表达式，其中 execution 就是一个切入点指示符。

execution 用于匹配执行方法的连接点，这是 Spring AOP 中最主要的切入点指示符。

execution 表达式的格式如下：

```
    execution(modifiers-pattern? ret-type-pattern
declaring-type-pattern? name-parrtern(param-pattern) throws-pattern?)
```

下面对 execution 表达式的格式进行解析。

① modifiers-pattern：指定方法的修饰符，支持通配符，该部分可省略。

② ret-type-pattern：指定方法的返回值类型，支持通配符，可以使用“*”通配符来匹配所有的返回值类型。

③ declaring-type-pattern：指定方法所属的类，支持通配符，该部分可省略。

④ name-parrtern：指定匹配方法名，支持通配符，可以使用“*”通配符来匹配所有的方法。

⑤ param-pattern：指定方法声明中的形参列表，支持通配符“*”和“..”，其中“*”代表一个任意类型的参数，而“..”代表 0 个或多个任意类型的参数。例如，()匹配一个不接受任何参数的方法，而（..）匹配一个接受任意数量参数的方法（0 个或多个），（*）匹配一个接受任意类型参数的方法。（*，String）匹配了一个接受两个参数的方法，第一个参数可以是任意类型，第二个参数必须是 String 类型。

⑥ throws-pattern：指定方法声明抛出的异常，支持通配符，该部分可省略。

下面给出常见切入点表达式的一些示例。

① 任意公共方法的执行：

```
execution(public * *(..))
```

② 任意一个以“set”开始的方法的执行：

```
execution(* set*(..))
```

③ AccountService 接口的任意方法的执行：

```
execution(* com.xyz.service.AccountService.*(..))
```

④ 定义在 service 包里的任意方法的执行：

```
execution(* com.xyz.service.*.*(..))
```

⑤ 定义在 service 包或者子包里的任意方法的执行：

```
execution(* com.xyz.service..*.*(..))
```

4.4.4　基于 XML 配置的 AOP 编程

使用 XML 配置方式时可通过 XML 文件来提供一些信息，如切面、切入点、增强处理，从

基于 XML 配置的 AOP 编程

而实现 AOP 编程。

（1）添加支持 AOP 编程的 JAR 包

需要在支持 Spring IoC 的核心 JAR 包的基础上，添加 aopalliance-1.0.jar、aspectjweaver.jar、spring-aop-5.0.2.RELEASE.jar、spring-aspects-5.0.2.RELEASE.jar，如图 4-1 所示。下面分别介绍这 4 个 JAR 包。

aopalliance-1.0.jar
aspectjweaver.jar
commons-logging-1.2.jar
spring-aop-5.0.2.RELEASE.jar
spring-aspects-5.0.2.RELEASE.jar
spring-beans-5.0.2.RELEASE.jar
spring-context-5.0.2.RELEASE.jar
spring-core-5.0.2.RELEASE.jar
spring-expression-5.0.2.RELEASE.jar

图 4-1　支持 AOP 编程新添加的 JAR 包

① aopalliance-1.0.jar：是 AOP 联盟提供的 API 包，里面包含了针对面向切面的接口。

② aspectjweaver.jar：是 Spring 的切入点表达式需要用的包。

③ spring-aop-5.0.2.RELEASE.jar：包含了使用 Spring AOP 特性时所需要的类，利用这个 JAR 文件可以使用基于 AOP 的 Spring 特性，如声明式的事务管理、日志系统引入等。

④ spring-aspects-5.0.2.RELEASE.jar：提供了对 AspectJ 的支持，以方便将面向切面的功能集成到 IDE 中。AspectJ 是一个面向切面的框架，它扩展了 Java 语言。AspectJ 定义了 AOP 语法，所以它有一个专门的编译器来生成 Java 字节规范的 class 文件。

（2）配置切面

既然使用 XML 配置方式实现 AOP 编程，就要在 XML 文件中定义切面、切入点、通知。Spring 的 XML 配置文件样例如下。

```
<beans xmlns="http://www.springframework.org/schema/beans"
    xmlns:xsi="http://www.w3.org/2001/XMLSchema-instance"
    xmlns:aop="http://www.springframework.org/schema/aop"
    xsi:schemaLocation="
         http://www.springframework.org/schema/beans
         http://www.springframework.org/schema/beans/spring-beans.xsd
         http://www.springframework.org/schema/aop
         http://www.springframework.org/schema/aop/spring-aop.xsd">
    <bean id="fooService" class="x.y.service.DefaultFooService"/>
    <bean id="profiler" class="x.y.SimpleProfiler"/>
    <aop:config>
         <aop:aspect ref="profiler">
             <aop:pointcut id="theExecutionOfSomeFooServiceMethod" expression=
                 "execution(* x.y.service.FooService.getFoo
             (String,int)) andargs(name, age)"/>
             <aop:around pointcut-ref="theExecutionOfSomeFooServiceMethod"
                 method="profile"/>
         </aop:aspect>
    </aop:config>
</beans>
```

关于头文件，因为要使用 aop 标签，则需在配置文件 beans 标签中引入 xmlns:aop 和 xsi: schemaLocation=" http://www.springframework.org/schema/aop http://www.springframework.org/schema/aop/spring-aop.xsd"。

关于配置文件中的 aop 相关标签介绍如下。

① aop:config：注明开始配置 AOP，是顶层的 AOP 配置标签，在 Spring 配置文件中的<beans>下可以包含多个子标签或属性，子标签包括<aop:pointcut>、<aop:advisor>、<aop:aspect>。在配置时，3 个子标签必须按照该顺序进行定义。

② aop:aspect：配置切面，可以包含子标签或属性，属性 ref 的值引用相关切面类的 Bean 的 id，order 可设置优先级（也可以不设置）。子标签包括<aop:pointcut>和不同通知类型。

可以配置以下几种通知：aop:before、aop:after、aop:after-returning、aop:after-throwing、aop:around。

③ aop:pointcut：配置切入点，属性 expression 的值是具体的表达式，id 是该 aop:pointcut 的唯一标识。

④ aop:before：定义前置通知，属性 pointcut-ref 为引用 aop:pointcut 定义的 id。属性 pointcut 可以直接写切入点表达式。属性 method 用于指定通知的方法名。

⑤ aop:after：定义最终通知，属性同 aop:before。

⑥ aop:around：定义环绕通知。

⑦ aop:after-returning：定义正常返回通知，属性同 aop:before。

⑧ aop:after-throwing：定义异常返回通知，属性同 aop:before。

下面通过案例演示基于 XML 配置文件的方式实现 AOP 编程。

【实例 4.1】基于 XML 配置文件的方式实现 AOP 编程案例 1。

具体源代码如下所示（代码详见 Spring5_Ch04_aopXML01\src\com\inspur\aspect\LogAspect.java；Spring5_Ch04_aopXML01\src\com\inspur\service\CustomerService.java；Spring5_Ch04_aopXML01\src\com\inspur\service\impl\CustomerServiceImpl.java；Spring5_Ch04_aopXML01\src\beans.xml；Spring5_Ch04_aopXML01\src\com\inspur\test\Test.java）。

（1）LogAspect.java

```
package com.inspur.aspect;
public class LogAspect {
    /**
     * 前置通知
     */
    public void beforePrintLog() {
        System.out.println("LogAspect 类 beforePrintLog 方法中的 log 日志打印
    开始了..");
    }
    /**
     * 正常返回通知
     */
    public void afterReturnPrintLog() {
        System.out.println("LogAspect 类 afterReturnPrintLog 方法中的 log 日志
```

```
        打印开始了..");
        }
        /**
         * 异常返回通知
         */
        public void afterThrowingPrintLog() {
            System.out.println("LogAspect类afterThrowingPrintLog方法中的log日志
        打印开始了..");
        }
        /**
         * 最终通知
         */
        public void afterPrintLog() {
            System.out.println("LogAspect类afterPrintLog方法中的log日志打印开始
              了..");
        }
    }
```

（2）CustomerService.java

```
    package com.inspur.service;
    public interface CustomerService {
        void saveCustomer();
    }
```

（3）CustomerServiceImpl.java

```
    package com.inspur.service.impl;
    import org.springframework.stereotype.Service;
    import com.inspur.service.CustomerService;
    @Service("customerService")
    public class CustomerServiceImpl implements CustomerService{
        public void saveCustomer() {
            System.out.println("保存客户信息");
            //int a=9/0;//该处代码为模拟异常，供测试返回异常通知使用
        }
    }
```

（4）beans.xml

```
    <?xml version="1.0" encoding="UTF-8"?>
    <beans xmlns="http://www.springframework.org/schema/beans"
        xmlns:xsi="http://www.w3.org/2001/XMLSchema-instance"
        xmlns:aop="http://www.springframework.org/schema/aop"
        xsi:schemaLocation="http://www.springframework.org/schema/beans
            https://www.springframework.org/schema/beans/spring-beans.xsd
            http://www.springframework.org/schema/aop
            https://www.springframework.org/schema/aop/spring-aop.xsd">
        <bean id="logAspect" class="com.inspur.aspect.LogAspect"/>
        <bean id="customerService"
            class="com.inspur.service.impl.CustomerServiceImpl">
        </bean>
        <aop:config>
          <aop:aspect id="myAspect" ref="logAspect">
              <!-- 配置前置通知：它永远在切入点方法之前执行 -->
```

```
        <aop:before method="beforePrintLog" pointcut="execution(*
           com.inspur.service.impl.*.*(..))"/>
        <!-- 配置正常返回通知：当切入点方法正常执行之后，该通知执行。它和异常返回通知
        只能有一个执行-->
        <aop:after-returning  method="afterReturnPrintLog"
      pointcut="execution(* com.inspur.service.impl.*.*(..))"/>
        <!-- 配置异常返回通知：当切入点方法执行产生异常后执行，它和正常返回通知是互斥的-->
        <aop:after-throwing  method="afterThrowingPrintLog"
      pointcut="execution(* com.inspur.service.impl.*.*(..))"/>
        <!-- 配置最终通知：无论切入点方法是否正常执行，它都会在其后面执行-->
        <aop:after method="afterPrintLog" pointcut=
      "execution(* com.inspur.service.impl.*.*(..))"/>
      </aop:aspect>
    </aop:config>
</beans>
```

（5）Test.java

```
package com.inspur.test;
import org.springframework.beans.factory.BeanFactory;
import org.springframework.context.support.ClassPathXmlApplicationContext;
import com.inspur.service.CustomerService;
public class Test {
    public static void main(String[] args) {
        BeanFactory factory=new ClassPathXmlApplicationContext("beans .xml");
        CustomerService customerService=(CustomerService)factory.
    getBean("customerService");
        customerService.saveCustomer();
    }
}
```

选中 Test.java 文件，单击鼠标右键，选择“Run As”菜单中的子菜单“1 Java Application”，运行程序后，在控制台上显示的结果如图 4-2 所示。

```
LogAspect类beforePrintLog方法中的log日志打印开始了..
保存客户信息
LogAspect类afterReturnPrintLog方法中的log日志打印开始了..
LogAspect类afterPrintLog方法中的log日志打印开始了..
```

图 4-2　基于 XML 配置的 AOP 编程案例 1

【程序解析】LogAspect.java 为切面类。该类中定义了前置通知、正常返回通知、异常返回通知、最终通知。这 4 个通知都是围绕着切入点方法的执行前后展开。前置通知永远在切入点方法之前执行。正常返回通知在切入点方法正常执行之后执行。异常返回通知是当切入点方法执行产生异常后执行，它和正常返回通知是互斥的。无论切入点方法是否正常执行，最终通知都会在其后执行。

在通知的配置中使用属性 pointcut 指定的切入点表达式，execution(* com.inspur.service.impl.*.*(..))是表示在 com.inspur.service.impl 包中任何类的任何方法中切入不同类型的通知。关于切入点表达式的内容可参见 4.4.3 节。

图 4-2 中验证了 4 种通知的执行顺序：before、after-returning、after-throwing、after 依次执行。

现在在 CustomerServiceImpl.java 的 public void saveCustomer 方法中添加代码“int a=9/0;”，

该处代码为模拟异常，供测试返回异常通知使用。因出现了异常，则不执行正常返回通知 afterReturnPrintLog()，转而执行异常返回通知 afterThrowingPrintLog()。运行结果如图 4-3 所示。

```
LogAspect类beforePrintLog方法中的log日志打印开始了..
保存客户信息
Exception in thread "main" LogAspect类afterThrowingPrintLog方法中的log日志打印开始了..
LogAspect类afterPrintLog方法中的log日志打印开始了..
java.lang.ArithmeticException: / by zero
```

图 4-3 验证 4 种通知的执行顺序

在实例 4.1 中演示了除环绕通知之外的 4 种通知类型。下面通过实例 4.2 演示环绕通知的应用。

【实例 4.2】基于 XML 配置文件的方式实现 AOP 编程案例 2。

具体源代码如下所示（代码详见 Spring5_Ch04_aopXML02\src\com\inspur\aspect\LogAspect.java；Spring5_Ch04_aopXML02\src\com\inspur\service\CustomerService.java；Spring5_Ch04_aopXML02\src\com\inspur\service\impl\CustomerServiceImpl.java；Spring5_Ch04_aopXML02\src\beans.xml；Spring5_Ch04_aopXML02\src\com\inspur\test\Test.java）。

其中 CustomerService.java、CustomerServiceImpl.java、Test.java 可参见实例 4.1。

（1）LogAspect.java

```
package com.inspur.aspect;
public class LogAspect {
    public void aroundPrintLog() {
        System.out.println("LogAspect 类 aroundPrintLog 方法中的 log 日志打印开始
    了...这是环绕通知.");
    }
}
```

（2）beans.xml

```
<?xml version="1.0" encoding="UTF-8"?>
<beans xmlns="http://www.springframework.org/schema/beans"
    xmlns:xsi="http://www.w3.org/2001/XMLSchema-instance"
    xmlns:aop="http://www.springframework.org/schema/aop"
    xsi:schemaLocation="http://www.springframework.org/schema/beans
         https://www.springframework.org/schema/beans/spring-beans.xsd
         http://www.springframework.org/schema/aop
         https://www.springframework.org/schema/aop/spring-aop.xsd">
    <bean id="logAspect" class="com.inspur.aspect.LogAspect"/>
    <bean id="customerService"
           class="com.inspur.service.impl.CustomerServiceImpl">
    </bean>
    <aop:config>
        <aop:aspect id="myAspect" ref="logAspect">
        <aop:around method="aroundPrintLog" pointcut="execution(*
            com.inspur.service.impl.*.*(..))"/>
      </aop:aspect>
    </aop:config>
```

选中 Test.java 文件，单击鼠标右键，选择“Run As”菜单中的子菜单“1 Java Application”，运行程序后，在控制台上显示的结果如图 4-4 所示。

```
LogAspect类aroundPrintLog方法中的log日志打印开始了...这是环绕通知.
```

图 4-4　基于 XML 配置的 AOP 编程案例 2

【程序解析】图 4-4 显示当配置了环绕通知后，执行切入点方法 saveCustomer 时，最终的结果是环绕通知的代码执行了，但切入点方法没有执行。下面分析一下原因：根据动态代理的代码分析，可以看到 invoke 方法中有明确调用切入点方法的代码，而 Spring 中的环绕通知目前没有调用切入点的方法。

那怎样解决呢？解决方法为要在环绕通知中明确调用切入点的方法。

Spring 框架提供了一个接口 ProceedingJoinPoint。该接口可以作为环绕通知的方法参数使用。在程序运行时 Spring 框架会注入该接口的实现类供用户使用。

该接口有方法 proceed，它相当于明确调用切入点的方法。

那么如何使用环绕通知呢？对比实例 4.1，其都是在 beans.xml 中配置通知，是 Spring 容器自动调用通知。而环绕通知在 Spring 中为用户提供了一种可以在代码中手动控制通知何时执行的方式。

下面就改进一下实例 4.2，具体参见实例 4.3。

【实例 4.3】基于 XML 配置文件的方式实现 AOP 编程案例 2 优化。

其他 Java 文件参见实例 4.2。

实例 4.3

LogAspect.java 修改如下：

```
package com.inspur.aspect;
import org.aspectj.lang.ProceedingJoinPoint;
public class LogAspect {
    /**
     * 环绕通知
     */
    public void aroundPrintLog(ProceedingJoinPoint pjp) {
        Object rtValue=null;
        System.out.println("LogAspect 类 aroundPrintLog 方法中的 log 日志打印开始
    了...这是前置通知.");
        try {
            Object[] objs=pjp.getArgs();
            rtValue=pjp.proceed(objs);
            System.out.println("LogAspect 类 aroundPrintLog 方法中的 log 日志打印
        开始了...这是正常返回通知.");
        } catch (Throwable e) {
            System.out.println("LogAspect 类 aroundPrintLog 方法中的 log 日志打印
        开始了...这是异常返回通知.");
            e.printStackTrace();
        }finally {
            System.out.println("LogAspect 类 aroundPrintLog 方法中的 log 日志打印
        开始了...这是最终通知.");
        }
    }
}
```

选中 Test.java 文件，单击鼠标右键，选择“Run As”菜单中的子菜单“1 Java Application”，

运行程序后，在控制台上显示的结果如图 4-5 所示。

```
LogAspect类aroundPrintLog方法中的log日志打印开始了...这是前置通知.
保存客户信息
LogAspect类aroundPrintLog方法中的log日志打印开始了...这是正常返回通知.
LogAspect类aroundPrintLog方法中的log日志打印开始了...这是最终通知.
```

图 4-5 基于 XML 配置的 AOP 编程案例 2（优化 1）

【程序解析】图 4-5 显示了正常执行切入点方法后，环绕通知的执行情况。结果中显示的前置通知、正常返回通知、最终通知的执行顺序是按照我们规定的顺序执行的。

现在在 CustomerServiceImpl.java 的 public void saveCustomer 方法中添加代码“int a=9/0; ”，该处代码为模拟异常，供测试返回异常通知使用。程序运行结果如图 4-6 所示。

```
LogAspect类aroundPrintLog方法中的log日志打印开始了...这是前置通知.
java.lang.ArithmeticException: / by zero
保存客户信息
LogAspect类aroundPrintLog方法中的log日志打印开始了...这是异常返回通知.
```

```
LogAspect类aroundPrintLog方法中的log日志打印开始了...这是最终通知.
```

图 4-6 基于 XML 配置的 AOP 编程案例 2（优化 2）

4.4.5 基于注解的 AOP 编程

基于注解的 AOP 编程

使用注解（Annotation）的方式已经逐渐成为当前的主流，下面我们利用上面的例子来说明如何用注解来开发 Spring AOP。

表 4-1 展示了 Spring 中的 AspectJ 注解。

表 4–1 Spring 中的 AspectJ 注解

注解	说明
@Aspect	配置切面
@Before	前置通知，在连接点方法前调用
@Around	环绕通知，它将覆盖原有方法，但允许用户手动定义通知执行的顺序
@After	最终通知，在连接点方法后调用
@AfterReturning	正常返回通知，在连接点方法执行并正常返回后调用，要求连接点方法在执行过程中没有发生异常
@AfterThrowing	异常返回通知，在连接点方法异常时调用

下面通过实例 4.4 演示注解@Aspect、@Before、@After、@AfterReturning、@AfterThrowing 的应用。

【实例 4.4】基于注解的 AOP 编程案例 1。

具体源代码如下所示（代码详见 Spring5_Ch04_aopAnnotation01\src\com\inspur\aspect\LogAspect.java；Spring5_Ch04_aopAnnotation01\src\com\inspur\service\CustomerService.java；Spring5_Ch04_aopAnnotation01\src\com\inspur\service\impl\CustomerServiceImpl.java；Spring5_Ch04_aopAnnotation01\src\beans.xml；Spring5_Ch04_aopAnnotation01\src\com\inspur\test\Test.java）。

其中 CustomerService.java、CustomerServiceImpl.java、Test.java 可参见实例 4.1。

（1）LogAspect.java

```
package com.inspur.aspect;
import org.aspectj.lang.annotation.After;
import org.aspectj.lang.annotation.AfterReturning;
import org.aspectj.lang.annotation.AfterThrowing;
import org.aspectj.lang.annotation.Aspect;
import org.aspectj.lang.annotation.Before;
import org.aspectj.lang.annotation.Pointcut;
import org.springframework.stereotype.Component;
@Component("logAspect")
@Aspect//把当前类配置成为一个切面
public class LogAspect {
    @Pointcut("execution(* com.inspur.service.impl.*.*(..))")
    private void ptr() {
    }
    /**
     * 前置通知
     */
    @Before("ptr()")
    public void beforePrintLog() {
        System.out.println("LogAspect 类 beforePrintLog 方法中的 log 日志打印开始了..");
    }
    /**
     * 正常返回通知
     */
    @AfterReturning("ptr()")
    public void afterReturnPrintLog() {
        System.out.println("LogAspect 类 afterReturnPrintLog 方法中的 log 日志打印开始了..");
    }
    /**
     * 异常返回通知
     */
    @AfterThrowing("ptr()")
    public void afterThrowingPrintLog() {
        System.out.println("LogAspect 类 afterThrowingPrintLog 方法中的 log 日志打印开始了..");
    }
    /**
     * 最终通知
     */
    @After("ptr()")
    public void afterPrintLog() {
        System.out.println("LogAspect 类 afterPrintLog 方法中的 log 日志打印开始了..");
    }
}
```

（2）beans.xml

```
<?xml version="1.0" encoding="UTF-8"?>
```

```
<beans xmlns="http://www.springframework.org/schema/beans"
    xmlns:xsi="http://www.w3.org/2001/XMLSchema-instance"
    xmlns:aop="http://www.springframework.org/schema/aop"
    xmlns:context="http://www.springframework.org/schema/context"
    xsi:schemaLocation="http://www.springframework.org/schema/beans
        https://www.springframework.org/schema/beans/spring-beans.xsd
        http://www.springframework.org/schema/aop
        https://www.springframework.org/schema/aop/spring-aop.xsd
        http://www.springframework.org/schema/context
        https://www.springframework.org/schema/context/spring-context.xsd">
    <!-- 告知 Spring 创建容器时自动扫描的包 -->
    <context:component-scan base-package="com.inspur"></context:
        component-scan>
    <!-- 启用 Spring 对注解 AOP 的支持 -->
    <aop:aspectj-autoproxy/>
</beans>
```

选中 Test.java 文件，单击鼠标右键，选择“Run As”菜单中的子菜单“1 Java Application”，运行程序后，在控制台上显示的结果如图 4-7 所示。

```
LogAspect类beforePrintLog方法中的log日志打印开始了..
保存客户信息
LogAspect类afterPrintLog方法中的log日志打印开始了..
LogAspect类afterReturnPrintLog方法中的log日志打印开始了..
```

图 4-7 基于注解的 AOP 编程案例 1 的运行结果 1

【程序解析】图 4-7 显示了正常执行切入点方法后，前置通知、正常返回通知、最终通知依次执行。

在 beans.xml 中配置代码“<context:component-scan base-package="com.inspur"></context:component-scan>”是告知 Spring 创建容器时自动扫描的包。代码“<aop:aspectj-autoproxy/>”启用了 Spring 对注解 AOP 的支持。这样就会自动扫描 com.inspur 包中 LogAspect.java 和 CustomerServiceImpl.java 的注解及解析注解。

在 LogAspect.java 中使用注解@Component("logAspect")，相当于 beans.xml 配置中的<bean id="logAspect" class="com.inspur.aspect.LogAspect"/>，完成建立 bean 的 id 为 LogAspect 的 LogAspect 类的对象。使用注解@Aspect 配置一个切面。使用注解@Pointcut 配置切入点，该切入点的 id 是方法名 ptr。使用注解@Before("ptr()")配置前置通知、@AfterReturning("ptr()")配置正常返回通知、@AfterThrowing("ptr()")配置异常返回通知、@After("ptr()")配置最终通知。以上注解完全相当于 beans.xml 配置文件中的<aop:aspect>的配置。

现在在 CustomerServiceImpl.java 的 public void saveCustomer 方法中添加代码“int a=9/0;”，该处代码为模拟异常，测试返回异常通知时使用。程序运行结果如图 4-8 所示。

```
LogAspect类beforePrintLog方法中的log日志打印开始了..
Exception in thread "main" java.lang.ArithmeticException: / by zero
```

```
保存客户信息
LogAspect类afterPrintLog方法中的log日志打印开始了..
LogAspect类afterThrowingPrintLog方法中的log日志打印开始了..
```

图 4-8 基于注解的 AOP 编程案例 1 的运行结果 2

因在实例 4.4 中没有验证@Around 注解，且还存在 beans.xml，所以并没有完全基于注解实现 AOP 编程。下面通过实例 4.5 演示全部基于注解实现 AOP 编程。

【实例 4.5】 基于注解的 AOP 编程案例 2。

实例 4.5

具体源代码如下所示（代码详见 Spring5_Ch04_aopAnnotation02\src\com\inspur\aspect\LogAspect.java；Spring5_Ch04_aopAnnotation02\src\com\inspur\service\CustomerService.java；Spring5_Ch04_aopAnnotation02\src\com\inspur\service\impl\CustomerServiceImpl.java；Spring5_Ch04_aopAnnotation02\src\config\SpringConfiguration.java；Spring5_Ch04_aopAnnotation02\src\com\inspur\test\Test.java）。

其中 CustomerService.java、CustomerServiceImpl.java 可参见实例 4.1。

（1）LogAspect.java

```
package com.inspur.aspect;
import org.aspectj.lang.ProceedingJoinPoint;
import org.aspectj.lang.annotation.After;
import org.aspectj.lang.annotation.AfterReturning;
import org.aspectj.lang.annotation.AfterThrowing;
import org.aspectj.lang.annotation.Around;
import org.aspectj.lang.annotation.Aspect;
import org.aspectj.lang.annotation.Before;
import org.aspectj.lang.annotation.Pointcut;
import org.springframework.stereotype.Component;
@Component("logAspect")
@Aspect//把当前类配置成为一个切面
public class LogAspect {
    @Around("execution(* com.inspur.service.impl.*.*(..))")
    public void aroundPrintLog(ProceedingJoinPoint pjp) {
        Object rtValue=null;
        System.out.println("LogAspect类aroundPrintLog方法中的log日志打印开始了...这是前置通知.");
        try {
            Object[] objs=pjp.getArgs();
            rtValue=pjp.proceed(objs);
            System.out.println("LogAspect类aroundPrintLog方法中的log日志打印开始了...这是正常返回通知.");
        } catch (Throwable e) {
            System.out.println("LogAspect类aroundPrintLog方法中的log日志打印开始了...这是异常返回通知.");
            e.printStackTrace();
        }finally {
            System.out.println("LogAspect类aroundPrintLog方法中的log日志打印开始了...这是最终通知.");
        }
    }
}
```

（2）SpringConfiguration.java

```
package config;
```

```
import org.springframework.context.annotation.ComponentScan;
import org.springframework.context.annotation.Configuration;
import org.springframework.context.annotation.EnableAspectJAutoProxy;
/**
 * Spring 的配置类，相当于 beans.xml
*/
@Configuration
@ComponentScan("com.inspur")
@EnableAspectJAutoProxy
public class SpringConfiguration {

}
```

（3）Test.java

```
package com.inspur.test;
import org.springframework.beans.factory.BeanFactory;
import org.springframework.context.annotation.AnnotationConfigApplicationContext;
import com.inspur.service.CustomerService;
import config.SpringConfiguration;
public class Test {
    public static void main(String[] args) {
        BeanFactory factory=new
    AnnotationConfigApplicationContext(SpringConfiguration.class);
        CustomerService customerService=
    (CustomerService)factory.getBean("customerService");
        customerService.saveCustomer();
    }
}
```

选中 Test.java 文件，单击鼠标右键，选择“Run As”菜单中的子菜单“1 Java Application”，运行程序后，在控制台上显示的结果如图 4-9 所示。

```
LogAspect类aroundPrintLog方法中的log日志打印开始了...这是前置通知.
保存客户信息
LogAspect类aroundPrintLog方法中的log日志打印开始了...这是正常返回通知.
LogAspect类aroundPrintLog方法中的log日志打印开始了...这是最终通知.
```

图 4-9　基于注解的 AOP 编程案例 2 的运行结果 1

【程序解析】图 4-9 显示执行环绕通知后，按照我们指定的前置通知、正常返回通知、最终通知依次执行，与使用 XML 配置结果相同。LogAspect.java 中使用@Around("execution(* com.inspur.service.impl.*.*(..))")完成注解环绕通知，其中的参数为切入点表达式。

Test.java 文件中代码“new AnnotationConfigApplicationContext(SpringConfiguration.class);”的作用是通过加载 SpringConfiguration 类创建 Spring 容器。

在 SpringConfiguration 类中使用了注解@Configuration 定义配置类，可替换 XML 配置文件。使用注解@ComponentScan("com.inspur")配置自动扫描的包，相当于 beans.xml 中的代码“<context:component-scan base-package="com.inspur"></context:component-scan>”。使用注解@EnableAspectJAutoProxy 启用 Spring 对注解 AOP 的支持，相当于 beans.xml 中的代码<aop:aspectj-autoproxy/>。

本章小结

通过本章的学习，读者首先需要掌握 AOP 的概念及相关术语，具体包括切面、连接点、通知、切入点、AOP 代理、织入、目标对象等，这些术语贯穿于整个 AOP 的应用中。在进行 AOP 编程前，读者需要理解 AOP 的两种代理机制：JDK 动态代理、CGLIB 代理。接着需要重点掌握 AOP 的编程，掌握 Spring 通知的类型，包括前置通知、正常返回通知、异常返回通知、最终通知和环绕通知。前 4 个通知都是 Spring 容器及切入点的定义，而环绕通知支持自定义前置通知、正常返回通知、异常返回通知、最终通知的执行顺序。最后需要掌握 AOP 编程的两种方式，第一种为基于 XML 配置的方式，即使用 Spring 配置文件来定义切入点和通知等；第二种为基于注解的方式，即使用注解来定义切入点和通知等。

习题

1. 简述 AOP 的概念。
2. 简述 AOP 的相关术语，包括切面、连接点、通知、切入点、AOP 代理、织入、目标对象等。
3. Spring 实现 AOP 的方式有哪些？

上机指导

1. 完成保存用户信息前自动切入 check()。
2. 分别使用基于 XML 配置的方式和基于注解的方式实现 AOP 编程。

05 第5章　事务管理

学习目标

- 了解事务的概念
- 了解事务的特性
- 了解 JDBC 事务管理的应用
- 掌握 Spring 编程式事务管理的应用方法
- 掌握 Spring 声明式事务管理的应用方法

5.1　事务的概念

事务的概念

数据库系统为了保证数据操作的完整性和一致性，引入了事务这个重要的概念。所谓事务，就是将一系列的数据库操作作为一个整体来执行。例如，对数据库存取，就是一组 SQL 指令，这一组 SQL 指令必须全部执行成功；如果其中一行 SQL 有错误，则先前执行过的 SQL 指令均会被撤销。

事务的特点是要么全部执行，要么全部不执行。下面以现实生活中银行转账操作作为例子，帮助读者更好地理解事务的概念。

例如，张三在银行办理转账业务，希望将 100 元人民币转到李四的账户上，这个转账操作必须执行以下两步才能完成。

① 从张三的账户上减去 100 元。

② 给李四的账户上增加 100 元。

以上两步必须保证同时执行成功。假设第一步完成后，由于通信故障等原因，导致第二步没有执行成功，则张三的账户扣除了 100 元，而李四的账户却没有增加相应的 100 元，那么银行的客户显然不会接受这种结果。这种情况下可以通过事务保证数据操作的完整性和一致性。

事务具备 ACID 特性，即原子性（Atomicity）、一致性（Consistency）、隔离性（Isolation）、持久性（Durability）。

① 原子性指整个数据库事务是不可分割的工作单位。只有数据库中所有的操作执行成功，才算整个事务成功；事务中任何一行 SQL 语句执行失败，已经执行成功的 SQL 语句也必须撤销，数据库状态应该退回到执行事务前的状态。

② 一致性是指数据库事务不能破坏关系数据的完整性以及业务逻辑上的一致性。例如上面提到的例子，银行转账事务，无论事务成功还是失败，都应保证转账结束后张三和李四的存款总额和转账前的存款总额保持一致。

③ 隔离性指的是在并发环境中，当不同的事务同时操纵相同的数据时，每个事务都有各自的完整数据空间。

④ 持久性指的是只要事务成功结束，它对数据库所做的更新就必须永久保存下来。即使发生系统崩溃，重新启动数据库系统后，数据库还能恢复到事务成功结束时的状态。

5.2　JDBC 事务管理

JDBC 事务管理

在 JDBC 中，Connection 提供了事务处理的方法，通过调用 setAutoCommit(false)可以设置手动提交事务；当事务完成后在一连串的 SQL 语句后面可用 commit()显式提交事务；如果在事务处理过程中发生异常，则可通过 rollback()进行事务回滚。除此之外，JDBC 中还引入了保存点（Savepoint）的概念，允许通过代码设置保存点并让事务回滚到指定的保存点，如图 5-1 所示。

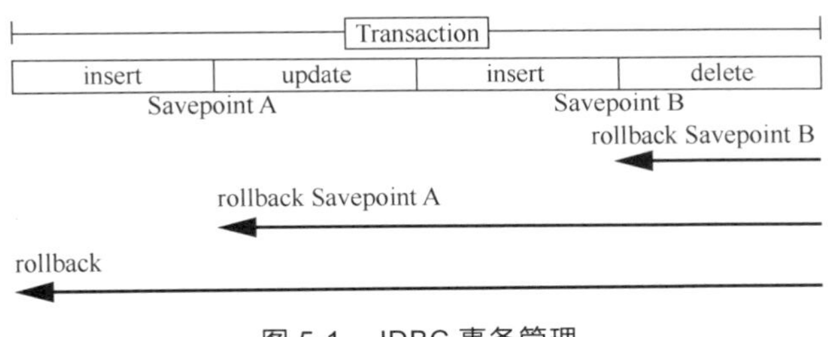

图 5-1　JDBC 事务管理

关键代码如下：

```
try{
    connection.setAutoCommit(false);         //设置手动提交事务
    …//一连串 SQL 操作
    connection.commit();                     //执行成功，提交所有变更
}catch(SQLException e){
    connection.rollback();                   //发生错误，撤销所有变更
}
```

下面通过实例演示 JDBC 事务管理。

【实例 5.1】演示 JDBC 事务管理的案例。

准备工作：建立 Java 工程并添加 JAR 包（如 Oracle 数据库的驱动包及 JDBC 事务包等）到编译路径上。在 Oracle 数据库中创建一个账号表 T_ACCOUNT 并插入两条数据，数据内容有账号名称和账号余额。

在数据库中创建账号表 T_ACCOUNT 及添加数据的 SQL 语句如下：

```
create table t_account(username varchar2(20),money number(10));
insert  into t_account(username,money) values('smith',1000);
insert  into t_account(username,money) values('tom',1000);
```

具体源代码如下所示（代码详见 Spring5_Ch05_Spring1\src\com\inspur\jdbctx\AccountDao.java；Spring5_Ch05_Spring1\src\com\inspur\jdbctx\AccountDaoImpl.java；Spring5_Ch05_Spring1\src\ApplicationContext.xml；Spring5_Ch05_Spring1\src\com\inspur\jdbctx\Client.java）。

（1）AccountDao.java

```
package com.inspur.jdbctx;
public interface AccountDao {
    //  增加账户金额
    public void addMoney(String username, Double money);
    // 减少账户金额
    public void subMoney(String username, Double money);
    // 查询账户金额
    public String queryAcountMoney(String username);
    // 转账
    public void tranfAccount(String sourceName, String targetName, Double
money);
}
```

（2）AccountDaoImpl.java

```
package com.inspur.jdbctx;
import java.sql.Connection;
import java.sql.PreparedStatement;
import java.sql.ResultSet;
import java.sql.SQLException;
import javax.sql.DataSource;
/**
* DAO 层实现类
* /
public class AccountDaoImpl implements AccountDao {
    private DataSource dataSource;
    private Connection conn;
    public DataSource getDataSource() {
        return dataSource;
    }
    public void setDataSource(DataSource dataSource) {
        this.dataSource = dataSource;
        try {
            conn=this.dataSource.getConnection();
        } catch (SQLException e) {
            e.printStackTrace();
        }
    }
/**
 * 增加账户金额
 */
    public void addMoney(String username, Double money) {
        PreparedStatement psmt=null;
```

```
        try {
            psmt=conn.prepareStatement("update  t_account set money=money+?
              where username=?");
            psmt.setDouble(1, money);
            psmt.setString(2, username);
            psmt.executeUpdate();
            System.out.println("增加账户金额成功");
        } catch (SQLException e) {
            try {
            conn.rollback();
            System.out.println("增加账户金额失败");
            } catch (SQLException e1) {
                e1.printStackTrace();
            }
            e.printStackTrace();
        }
    }
  /**
    * 减少账户金额
    */
   public void subMoney(String username, Double money) {
       PreparedStatement psmt=null;
       try {
           psmt=conn.prepareStatement("update  t_account set money=money-?
              where username=?");
           psmt.setDouble(1, money);
           psmt.setString(2, username);
           psmt.executeUpdate();
           System.out.println("减少账户金额成功");
       } catch (SQLException e) {
           try {
               conn.rollback();
               System.out.println("减少账户金额失败");
           } catch (SQLException e1) {
               e1.printStackTrace();
           }
           e.printStackTrace();
       }
   }
   /**
    * 账号转账
    */
   public void tranfAccount(String sourceName, String targetName, Double
money) {
       try {
           //设置为手动提交事务
           conn.setAutoCommit(false);
           subMoney(sourceName,money);
           Integer.parseInt("asfdas");  //模拟出现异常
           addMoney(targetName,money);
           System.out.println("转账成功");
           //提交事务
           conn.commit();
       } catch (Exception e) {
```

```
                e.printStackTrace();
                try {
                //回滚事务
                conn.rollback();
                System.out.println("转账失败");
                } catch (SQLException e1) {
                    e1.printStackTrace();
                }
            }
        }
    }
```

（3）ApplicationContext.xml

```
<?xml version="1.0" encoding="GB18030"?>
<beans xmlns="http://www.springframework.org/schema/beans"
    xmlns:xsi="http://www.w3.org/2001/XMLSchema-instance"
    xsi:schemaLocation="http://www.springframework.org/schema/beans
    http://www.springframework.org/schema/beans/spring-beans.xsd">
 <bean id="dataSource" class="org.springframework.jdbc.datasource.
   DriverManagerDataSource">
   <property name="driverClassName" value="oracle.jdbc.driver.OracleDriver">
 </property>
   <property name="url" value="jdbc:oracle:thin:@localhost:1521:xe"></property>
   <property name="username" value="scott"></property>
   <property name="password" value="tiger"></property>
 </bean>
 <bean id="accountDao" class="com.inspur.jdbctx.AccountDaoImpl">
   <property name="dataSource" ref="dataSource"></property>
 </bean>
</beans>
```

（4）Client.java

```
package com.inspur.jdbctx;
import org.springframework.context.ApplicationContext;
import org.springframework.context.support.ClassPathXmlApplicationContext;
public class Client {
    public static void main(String[] args) {
        ApplicationContext ctx=new
           ClassPathXmlApplicationContext("ApplicationContext.xml");
        AccountDao accountDao=(AccountDao)ctx.getBean("accountDao");
        accountDao.tranfAccount("smith", "tom",1000);
        System.out.println(accountDao.queryAcountMoney("smith"));
    }
}
```

选中该 Client.java 文件，单击鼠标右键，选择“Run As”菜单中的子菜单“1 Java Application”，运行程序后，在控制台上显示的结果如图 5-2 所示。

```
java.lang.NumberFormatException: For input string: "asfdas"
	at java.lang.NumberFormatException.forInputString(Unknown Source)
	at java.lang.Integer.parseInt(Unknown Source)
	at java.lang.Integer.parseInt(Unknown Source)
	at com.inspur.jdbctx.AccountDaoImpl.tranfAccount(AccountDaoImpl.java:99)
	at com.inspur.jdbctx.Client.main(Client.java:17)
```

图 5-2　验证 JDBC 事务管理运行结果

【案例解析】

（1）在 ApplicationContext.xml 中进行数据源的配置：DriverManagerDataSource 是 Spring 提供的一个简单的数据源实现类，它位于 org.springframework.jdbc.datasource 包中。这个类实现了 javax.sql.DataSource 接口，但它并没有提供池化连接的机制，每次调用 getConnection()获取新连接时，都只会简单地创建一个新的连接。这个数据源类比较适合在单元测试或简单的独立应用中使用，因为它不需要额外的依赖类。

（2）通过运行程序会发现：如果在 Client.java 的 main 方法中单独执行增加账户金额的 addMoney 方法或减少账户金额的 subMoney 方法，程序能够执行成功。如果执行转账操作的 tranfAccount 方法，因为转账是一个整体，在该方法中执行的操作是在一个事务中，所以若在调用 addMoney 方法和 subMoney 方法的代码中间有一行异常代码 "Integer.parseInt("asfdas");"，则控制台会报异常，并且 addMoney 方法和 subMoney 方法都不会执行成功，执行结果如图 5-2 所示。

如果把这行异常代码去掉，则转账成功，如图 5-3 所示。

```
减少账户金额成功
增加账户金额成功
转账成功
```

图 5-3　JDBC 事务执行成功结果

5.3　Spring 事务管理

Spring 框架提供了极其强大且又简便的事务处理功能。Spring 将所有的事务管理都抽象为 PlatformTransactionManager、TransactionStatus 和 TransactionDefinition 这 3 个接口，而无论其底层关联的具体事务究竟是 JDBC 事务、JTA（Java Transaction API，Java 事务 API）事务，还是 ORM 框架自定义的事务。

PlatformTransactionManager 接口定义了事务管理器，所有与事务相关的操作都由它管理；TransactionStatus 接口代表事务本身，它提供了简单的控制事务执行和查询事务状态的方法；TransactionDefinition 接口定义了事务的规则（如事务的隔离级别、传播行为、事务超时、只读状态），在启动事务时，事务管理器会根据 TransactionDefinition 来启动合适的事务。

Spring 进行事务处理的核心是 PlatformTransactionManager 抽象接口，由其完成获得事务、提交事务、回滚事务操作。

在 Spring 中实现事务管理有两种方式。一种是传统的编程式事务管理，也就是程序员编写程序代码实现事务的管理，具体包括定义事务的开始、在程序异常时进行事务的回滚及程序正常执行后的事务提交。另一种则是基于 AOP 技术实现的声明式事务管理，事务管理本身是一项共有的系统级服务功能，完全可以将事务管理抽象成一个事务切面，程序员不再关心事务管理的问题，而是把主要精力放在核心业务逻辑代码的编写上，然后在需要进行事务管理的方法上切入事务切面，使之具有事务管理的功能，达到事务管理的目的。

5.3.1 编程式事务管理

Spring 框架进行事务处理，其核心便是 PlatformTransactionManager 抽象接口。PlatformTransactionManager 是一个事务管理器接口，它只定义了 3 种方法：getTransaction 用来获取事务的状态；commit 用来提交事务；rollback 用来回滚事务。事务管理器的实现类有多种，根据具体的持久层框架的不同而不同，实现类有 DataSourceTransactionManager、HibernateTransactionManager、JdoTransactionManager 等。

Spring 编程式事务可以使用 PlatformTransactionManager 直接管理事务。首先，我们用 Spring 容器创建 PlatformTransaction 对象；然后，使用 TransactionDefinition 定义事务规则，使用 PlatformTransaction 对象根据事务规则获取一个事务 TransactionStatus 对象；最后 PlatformTransaction 对象可以根据业务逻辑处理提交事务或回滚事务。

总结 Spring 编程式事务一般流程如下。

（1）在 Spring 配置文件中声明数据源。

（2）在 Spring 配置文件中声明一个事务管理类，例如，DataSourceTransactionManager、HibernateTransactionManger、JTATransactionManager 等。

（3）在代码中加入事务处理代码。

事务处理关键代码如下。

```
//初始化事务
TransactionDefinition
tdefinition=new DefaultTransactionDefinition();
//获取当前事务
TransactionStatus ts = transactionManager.getTransaction(td);
try{
//逻辑处理语句
  transactionManager.commit(ts);
} catch(Exception e){
     transactionManager.rollback(ts);
}
```

下面通过实例演示编程式事务管理。

【实例 5.2】 演示编程式事务管理的案例。

准备工作：见实例 5.1 的准备工作。

具体源代码如下所示（代码详见 Spring5_Ch05_Spring2\src\com\inspur\jdbctx\AccountDao.java；Spring5_Ch05_Spring2\src\com\inspur\jdbctx\AccountDaoImpl.java；Spring5_Ch05_Spring2\src\ApplicationContext.xml；Spring5_Ch05_Spring2\src\com\inspur\jdbctx\Client.java）。

（1）AccountDao.java

参见实例 5.1 中的 AccountDao.java。

（2）AccountDaoImpl.java

```
package com.inspur.jdbctx;
import java.sql.Connection;
import java.sql.PreparedStatement;
```

```
import java.sql.ResultSet;
import java.sql.SQLException;
import javax.sql.DataSource;
import org.springframework.jdbc.core.JdbcTemplate;
import org.springframework.jdbc.datasource.DataSourceTransactionManager;
import org.springframework.transaction.TransactionStatus;
import org.springframework.transaction.support.DefaultTransactionDefinition;
public class AccountDaoImpl implements AccountDao {
    private DataSourceTransactionManager transactionManager;
    private JdbcTemplate jdbcTemplate;
    public DataSourceTransactionManager getTransactionManager() {
        return transactionManager;
    }
    public void setTransactionManager(
        DataSourceTransactionManager transactionManager) {
        this.transactionManager = transactionManager;
    }
    public JdbcTemplate getJdbcTemplate() {
        return jdbcTemplate;
    }
    public void setJdbcTemplate(JdbcTemplate jdbcTemplate) {
        this.jdbcTemplate = jdbcTemplate;
    }
    /**
     *增加账户金额
     */
    public void addMoney(String username, Double money) throws Exception {
        try {
            jdbcTemplate.update("update  t_account set money=money+? where
        username=?",new Object[]{new Double(money),username});
        } catch (Exception e) {
            throw new Exception();
        }
    }
    /**
     *减少账户金额
     */
    public void subMoney(String username, Double money) throws Exception {
        try {
            jdbcTemplate.update("update  t_account set money=money-? where
        username=?",new Object[]{new Double(money),username});
        } catch (Exception e) {
            throw new Exception();
        }
    }
    /**
     *转账
     */
    public void tranfAccount(String sourceName, String targetName, Double
money) {
        DefaultTransactionDefinition tdefinition=new DefaultTransactionDefinition();
        tdefinition.setPropagationBehavior(DefaultTransactionDefinition.
        PROPAGATION_REQUIRED);
        TransactionStatus ts=transactionManager.getTransaction(tdefinition);
```

```
            try {
                subMoney(sourceName,money);
                System.out.println("转出中......");
                Integer.parseInt("asfdas");
                addMoney(targetName,money);
                System.out.println("转入中......");
                transactionManager.commit(ts);
                System.out.println("转账成功");
            } catch (Exception e) {
                e.printStackTrace();
                transactionManager.rollback(ts);
                System.out.println("转账失败");
            }
        }
    }
```

（3）ApplicationContext.xml

```
<?xml version="1.0" encoding="GB18030"?>
<beans xmlns="http://www.springframework.org/schema/beans"
   xmlns:xsi="http://www.w3.org/2001/XMLSchema-instance"
   xsi:schemaLocation="http://www.springframework.org/schema/beans
   http://www.springframework.org/schema/beans/spring-beans.xsd">
<bean id="dataSource"
   class="org.springframework.jdbc.datasource.DriverManagerDataSource">
   <property name="driverClassName" value="oracle.jdbc.driver.OracleDriver">
   </property>
   <property name="url" value="jdbc:oracle:thin:@localhost:1521:xe"></property>
   <property name="username" value="scott"></property>
   <property name="password" value="tiger"></property>
</bean>
<!-- 配置 JDBC 模板 -->
<bean id="jdbcTemplate" class="org.springframework.jdbc.core.JdbcTemplate">
   <!-- 默认必须使用数据源 -->
   <property name="dataSource" ref="dataSource"/>
</bean>
 <!--配置事务管理器-->
<bean id="transactionManager"
    class="org.springframework.jdbc.datasource.DataSourceTransactionManager">
    <property name="dataSource" ref="dataSource"></property>
</bean>
<bean id="accountDao" class="com.inspur.jdbctx.AccountDaoImpl">
   <property name="transactionManager" ref="transactionManager"></property>
   <!-- 将 jdbcTemplate 注入 AccountDao 实例中 -->
   <property name="jdbcTemplate" ref="jdbcTemplate"></property>
</bean>
</beans>
```

（4）Client.java

参见实例 5.1 的 Client.java。

选中 Client.java 文件，单击鼠标右键，选择"Run As"菜单中的子菜单"1 Java Application"，运行程序后，在控制台上显示的结果如图 5-4 所示。

```
转出中.........
java.lang.NumberFormatException: For input string: "asfdas"
	at java.lang.NumberFormatException.forInputString(Unknown Source)
	at java.lang.Integer.parseInt(Unknown Source)
	at java.lang.Integer.parseInt(Unknown Source)
	at com.inspur.jdbctx.AccountDaoImpl.tranfAccount(AccountDaoImpl.java:137)
	at com.inspur.jdbctx.Client.main(Client.java:21)
转账失败
```

图 5-4 验证编程式事务管理运行结果

【案例解析】

（1）在 AccountDaoImpl.java 中的转账方法中创建了 org.springframework.transaction.support.DefaultTransactionDefinition 类的实例。DefaultTransactionDefinition 是 TransactionDefinition 接口的默认实现类，它提供了各事务属性的默认值，并且通过它可以更改这些值。这些默认值包括 propagationBehavior=PROPAGATION_REQUIRED、isolationLevel=ISOLATION_DEFAULT、timeout=TIMEOUT_DEFAULT、readonly=false。

（2）通过运行程序同样会发现：如果在 Client.java 的 main 方法中单独执行增加账户金额的 addMoney 方法或减少账户金额的 subMoney 方法，程序能够执行成功。如果执行转账操作的 tranfAccount 方法，因为转账是一个整体，在该方法中执行的操作是在一个事务中，所以若在调用 addMoney 方法和 subMoney 方法的代码中间有一行异常代码"Integer.parseInt("asfdas");"，则控制台会报异常，并且 addMoney 方法和 subMoney 方法都不会执行成功。只有把这行异常代码去掉，转账才能成功。

5.3.2 声明式事务管理

声明式事务管理

Spring 为声明式事务提供了简单且又强大的支持，所谓声明式事务，是指在 Spring 的配置文件中使用相应的标签对事务进行配置，这样做的好处是 Spring 可以帮助我们管理事务。例如，什么时候提交事务、什么时候回滚事务等。

从开发效率与易维护的角度来看，Spring 声明式事务管理是实际开发中比较常用的。

Spring 声明式事务的实现方式有以下 4 种。

① 基于 TransactionInterceptor 的声明式事务管理。

② 基于 TransactionProxyFactoryBean 的声明式事务管理。

③ 基于 tx 命名空间的声明式事务管理。

④ 基于@Transactional 的声明式事务管理。

下面讲解常用的方式——基于 tx 命名空间的声明式事务管理。

基于 tx 命名空间的声明式事务管理是通过在配置文件中配置事务规则的相关声明来实现的。其配置步骤如下。

① 在 Spring 配置文件中启用 tx 命名空间。

```
xmlns:tx=http://www.springframework.org/schema/tx
xsi:schemaLocation="http://www.springframework.org/schema/tx
```

```
http://www.springframework.org/schema/tx/spring-tx-5.0.xsd
```

② 在 xml 中配置通知、切入点以及 Advisor（通知器）。

tx 命名空间下提供了<tx:advice>来配置事务的通知（增强处理）。当使用<tx:advice>配置了事务的增强处理后，就可以通过编写的 AOP 配置，让 Spring 自动对目标生成代理。

配置<tx:advice>时，通常需要制定 id 和 transaction-manager 属性，其中 id 属性是配置文件的唯一标识，transaction-manager 属性用于指定事务管理器。还需要配置一个<tx:attributes>子标签，该子标签可通过配置多个<tx:method>子标签来配置执行事务的细节。

此外，还需定义<aop:config>标签，确保有 tx:advice 切面定义事务增强处理能力在合适的点被织入，如图 5-5 所示。

```
<tx:advice id="testAdvice" transaction-manager="transactionManager">
  <tx:attributes>
   <tx:method name="*" propagation="REQUIRED" isolation="READ_COMMITTED" />
  </tx:attributes>
</tx:advice>
<!-- 配置切入点和advisor -->
<aop:config>
 <aop:pointcut id="myPointcut" expression="execution(* com.inspur.aopspringtx.dao.*.tranfAccount(..))"/>
 <aop:advisor advice-ref="testAdvice" pointcut-ref="myPointcut"/>
</aop:config>
```

图 5-5　声明式事务 xml 中配置

图 5-5 中，tx 命名空间为声明在 Spring context 中的那些 Beans 提供声明式事务处理的支持，tx:advice 声明具备事务性质的 advice，tx:attributes 为一个或多个方法声明 Spring 事务的规则，tx:method 为一个给定的方法签名描述事务规则（事务的属性和具体对哪些方法执行事务），属性 transaction-manager 的默认值是 transactionManager，如果事务管理器 Bean 的名字是 transactionManager，则可以省略 transaction-manager 属性的设定。

下面通过实例演示声明式事务管理。

【实例 5.3】演示声明式事务管理的案例。

准备工作：见实例 5.1 的准备工作。

具体源代码如下所示（代码详见 Spring5_Ch05_Spring3\src\com\inspur\jdbctx\AccountDao.java；Spring5_Ch05_Spring3\src\com\inspur\jdbctx\AccountDaoImpl.java；Spring5_Ch05_Spring3\src\ApplicationContext.xml；Spring5_Ch05_Spring3\src\com\inspur\jdbctx\Client.java）。

（1）AccountDao.java

参见实例 5.1 中的 AccountDao.java。

（2）AccountDaoImpl.java

```
package com.inspur.jdbctx;
import java.sql.Connection;
import java.sql.PreparedStatement;
import java.sql.ResultSet;
import java.sql.SQLException;
import javax.sql.DataSource;
import org.springframework.jdbc.core.JdbcTemplate;
import org.springframework.jdbc.datasource.DataSourceTransactionManager;
```

```
import org.springframework.transaction.TransactionStatus;
import org.springframework.transaction.support.DefaultTransactionDefinition;
public class AccountDaoImpl implements AccountDao {
    private JdbcTemplate jdbcTemplate;
    /**
     * 增加账户金额
     */
    public void addMoney(String username, double money) throws Exception {
        PreparedStatement psmt=null;
        try {
            jdbcTemplate.update("update  t_account set money=money+? where
               username=?", money,username);
        } catch (Exception e) {
            throw new Exception();
        }
    }
    //减少账户金额
    public void subMoney(String username, double money) throws Exception {
        try {
            jdbcTemplate.update("update  t_account set money=money-? where
               username=?",money,username);
        } catch (Exception e) {
            throw new Exception();
        }
    }
    /**
     * 转账
     */
    public void tranfAccountMoney(String sourceName, String targetName, double
money) {
        try {
            subMoney(sourceName,money);
            Integer.parseInt("asfdas");  //模拟出现异常
            addMoney(targetName,money);
        } catch (Exception e) {
            e.printStackTrace();
            System.out.println("转账失败");
        }
    }
}
```

（3）ApplicationContext.xml

```
<?xml version="1.0" encoding="GB18030"?>
<beans xmlns="http://www.springframework.org/schema/beans"
    xmlns:xsi="http://www.w3.org/2001/XMLSchema-instance"
    xmlns:aop="http://www.springframework.org/schema/aop"
    xmlns:tx="http://www.springframework.org/schema/tx"
    xsi:schemaLocation="http://www.springframework.org/schema/beans
          http://www.springframework.org/schema/beans/spring-beans.xsd
          http://www.springframework.org/schema/aop
          http://www.springframework.org/schema/aop/spring-aop.xsd
          http://www.springframework.org/schema/tx
          http://www.springframework.org/schema/tx/spring-tx.xsd">
<bean id="dataSource"
```

```
        class="org.springframework.jdbc.datasource.DriverManagerDataSource">
        <property name="driverClassName" value="oracle.jdbc.driver.OracleDriver">
    </property>
        <property name="url" value="jdbc:oracle:thin:@localhost:1521:xe"></property>
        <property name="username" value="scott"></property>
        <property name="password" value="tiger"></property>
     </bean>
     <!--配置事务管理器-->
    <bean id="transactionManager"
         class="org.springframework.jdbc.datasource.DataSourceTransactionManager">
    <property name="dataSource" ref="dataSource"></property>
    </bean>
         <!--配置事务的通知-->
         <tx:advice id="transaction" transaction-manager="transactionManager">
             <!-- 配置事务的属性为每类方法配置各自的事务 -->
             <tx:attributes>
                 <tx:method name="*" propagation="REQUIRED"
                    isolation="READ_COMMITTED" />
             </tx:attributes>
         </tx:advice>
         <!--  配置 AOP: 配置切入点和 advisor-->
         <aop:config>
             <!-- 配置切入点表达式-->
            <aop:pointcut id="txPointcut" expression="execution(* *Money(..))" />
             <!--建立切入点表达式和事务通知的对应关系 -->
             <aop:advisor advice-ref="transaction" pointcut-ref="txPointcut" />
         </aop:config>
         <!-- 配置业务对象 -->
     <bean id="accountDao" class="com.inspur.jdbctx.AccountDaoImpl">
        <!-- 将 jdbcTemplate 注入 AccountDao 实例中 -->
        <property name="jdbcTemplate" ref="jdbcTemplate"></property>
    </beans>
```

（4）Client.java

参见实例 5.1 的 Client.java。

选中 Client.java 文件，单击鼠标右键，选择“Run As”菜单中的子菜单“1 Java Application”，运行程序后，在控制台上显示的结果如图 5-6 所示。

```
转账失败
java.lang.NumberFormatException: For input string: "asfdas"
	at java.lang.NumberFormatException.forInputString(Unknown Source)
	at java.lang.Integer.parseInt(Unknown Source)
	at java.lang.Integer.parseInt(Unknown Source)
	at com.inspur.jdbctx.AccountDaoImpl.tranfAccountMoney(AccountDaoImpl.java:138)
	at sun.reflect.NativeMethodAccessorImpl.invoke0(Native Method)
	at sun.reflect.NativeMethodAccessorImpl.invoke(Unknown Source)
```

图 5-6 验证声明式事务管理运行结果

【案例解析】

与前两个例子一样，通过运行程序可以发现：如果在 AccountDaoImpl.java 的转账 tranfAccountMoney 方法中去掉那行异常代码“Integer.parseInt("asfdas");”，程序就能正确执行且转账成功，否则就会转账失败。

Spring 中基于 XML 的声明式事务控制配置步骤总结如下。

（1）配置事务管理器。

在 Bean 文件中配置事务管理器，配置时需要注入数据源。

（2）配置事务的通知。

① 需要先导入 tx 命名空间。

② 使用<tx:advice>声明事务通知，需要指定 id 属性，以便 AOP 把通知和切入点关联起来；还需要指定 transaction-manager 属性，其值为 Bean 配置文件中事务管理器的 id 属性值。

（3）配置 AOP 中的通用切入点表达式。

在<aop:config>下，使用<aop:pointcut>声明切入点，其 expression 属性指定切入点表达式，还需要指定 id 属性。

（4）建立事务通知和切入点表达式的对应关系。

在<aop:config>下，使用<aop:advisor>声明一个增强器，将事务通知和切入点关联起来，使用 advice-ref 属性指定事务通知，使用 pointcut-ref 属性指定切入点。

（5）配置事务的属性。在事务的通知<tx:advice>标签的内部配置。

① 在<tx:advice>下声明<tx:attributes>，用于指定事务属性。

② 在<tx:attributes>下可以使用多个<tx:method>指定多种事务属性。

5.4　Spring 事务的传播方式和隔离级别

Spring 事务的传播方式和隔离级别

下面对 Spring 事务的传播方式和隔离级别进行详细介绍。

5.4.1　传播方式

事务的传播行为用于指定在多个事务方法间调用时，事务在这些方法间传播。Spring 支持以下 7 种传播方式。

（1）PROPAGATION_REQUIRED

表示业务逻辑方法需要在一个事务中运行，如果该方法在运行时，已经处于一个事务中，则直接加入该事务中，否则自己创建一个新的事务。即如果存在一个事务，则支持当前事务；如果没有事务，则开启一个新的事务（在实际开发中常使用该传播方式）。

（2）PROPAGATION_SUPPORTS

表示业务逻辑方法如果在某个事务范围内被调用，则该方法直接成为当前事务的一部分。如果该方法在事务范围外被调用，则该方法在无事务的环境下执行。即如果存在一个事务，则支持当前事务；如果没有事务，则非事务地执行。

（3）PROPAGATION_MANDATORY

表示业务逻辑方法只能在一个已经存在的事务中执行，该方法不能创建自己的事务，如果该方法在没有事务的环境下被调用，容器就会抛出事务不存在的异常。即如果已经存在一个事务，则支持当前事务；如果没有一个活动的事务，则抛出异常。

（4）PROPAGATION_REQUIRES_NEW

表示不管当前是否有事务存在，该业务逻辑方法都会为自己创建一个全新的事务。如果该方法已经运行在一个事务中，则原有事务会被挂起，新的事务会被创建，直到该方法执行结束后新事务才算结束，原先的事务再恢复执行。即总是开启一个新的事务。如果一个事务已经存在，则将这个存在的事务挂起，新事务运行完毕后，再接着运行被挂起的事务。

（5）PROPAGATION_NOT_SUPPORTED

表示业务逻辑方法不需要事务。如果该方法目前没有关联到某个事务，容器不会为它创建事务。如果该方法在一个事务中被调用，则该事务会被挂起，在方法调用结束后原先的事务才会恢复执行。即总是非事务地执行，并挂起任何存在的事务，当前方法运行完毕后，被挂起的事务才恢复执行。

（6）PROPAGATION_NEVER

表示业务逻辑方法绝对不能在事务范围内执行。如果该方法在某个事务中执行，容器会抛出异常，只在没有关联到任何事务时，该方法才能正常执行。即总是非事务地执行，如果存在一个活动事务，则抛出异常。

（7）PROPAGATION_NESTED

表示如果一个活动的事务存在，则该业务逻辑方法会运行在一个嵌套的事务中，如果没有活动事务，则按 REQUIRED 属性执行。它使用了一个单独的事务，这个事务拥有多个可以回滚的保存点，内部事务的回滚不会对外部事务造成影响，它只对 DataSourceTransactionManager 事务管理器生效。即如果一个活动的事务存在，则运行在一个嵌套的事务中。如果没有活动的事务，则按 TransactionDefinition.PROPAGATION_REQUIRED 属性执行。Spring 事务前 6 种传播方式如图 5-7 所示。

事务播放属性	t1（相当于UserManagerImpl）	t2（相当于LogImpl）
REQUIRED	无 t1	t2（t2自己会开启一个事务） t1（如果t1开启了事务，则t2直接使用t1开启的事务）
REQUIRES_NEW	无 t1	t2 t2（t2会自己单独开启一个新事务，不会使用t1的事务）
SUPPORTS	无 t1	无 t1（如果有事务，则直接使用）
MANDATORY	无 t1	抛出异常（如果没事务就抛出异常） t1（如果有事务，则直接使用）
NOT_SUPPORTED	无 t1	无 无（如果t1有事务，则t2也不用事务，不支持事务）
NEVER	无 t1	无 抛出异常

图 5-7 Spring 事务管理——前 6 种事务传播方式

对 Spring 的 7 种传播方式的传播特性总结如下。

对基于方法级别的事务管理而言，方法开始执行时创建事务，方法运行过程中若出现异常则进行事务回滚，方法如果正常执行则进行事务的提交。因而事务管理的主要任务就是事务的创建、事务的回滚与事务的提交，其中是否需要创建事务及如何创建事务是由事务传播行为控制的，通常数据的读取不需要事务管理，或者也可为其指定只读事务，而对于插入、修改与删除数据的方法来说，就有必要进行事务管理了。在未指定事务传播行为时，自 Spring 2.x 起就会启用默认的 REQUIRED。

5.4.2 隔离级别

事务处理时如果不考虑隔离性会引发脏读、不可重复读、虚幻读等安全性问题。脏读是指一个事务读到了另一个事务未提交的数据；不可重复读是指一个事务读到了另一个事务已经提交的 update 的数据，导致多次查询结果不一致；虚幻读指的是一个事务读到了另一个事务已经提交的 insert 的数据，导致多次查询结果不一致。

事务的隔离级别有以下 5 种。

（1）ISOLATION_DEFAULT

这是一个 PlatfromTransactionManager 默认的隔离级别，使用数据库默认的事务隔离级别。

以下 4 个与 JDBC 的隔离级别相对应。

（2）ISOLATION_READ_UNCOMMITTED

这是事务最低的隔离级别，它允许另外一个事务可以看到这个事务未提交的数据。这种隔离级别会产生脏读、不可重复读和虚幻读。

（3）ISOLATION_READ_COMMITTED

这种隔离级别保证一个事务修改的数据提交后才能被另外一个事务读取。另外一个事务不能读取该事务未提交的数据。

（4）ISOLATION_REPEATABLE_READ

这种事务隔离级别可以防止脏读、不可重复读，但可能出现虚幻读。它除了保证一个事务不能读取另一个事务未提交的数据外，还可以避免不可重复读。

（5）ISOLATION_SERIALIZABLE

这是花费代价最高但也是最可靠的事务隔离级别。事务被处理为顺序执行。除了防止脏读、不可重复读外，还避免了虚幻读。但是其并发性最差。

本章小结

本章首先讲解了事务的概念，事务的特性，JDBC 事务管理的步骤和操作过程，以及 Spring 事务管理的两种实现方式：编程式事务管理和声明式事务管理。进行事务处理时，编程式事务需要在代码中直接加入处理事务的逻辑，声明式事务的做法则是通过 AOP 面向切面编程的思想直接在配置文件中配置事务。本章最后通过案例对 Spring 事务处理的应用进行了详细讲解。通过本章的学习，读者可以对 Spring 的事务管理知识有一定的了解，并能够掌握 Spring 声明式事务管理的应用方法。

习题

1. 什么是事务？
2. 简述 Spring 事务管理的实现方式。
3. 简述 Spring 事务管理涉及的接口。
4. 简述 Spring 声明式事务如何配置。
5. 简述 Spring 事务的传播性。

上机指导

编写一个订单的案例来演示 Spring 中的事务管理，客户表和订单表的关系模型如图 5-8 所示。

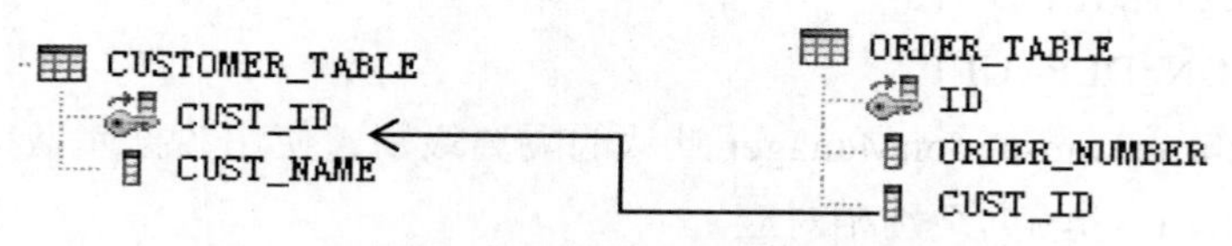

图 5-8　客户表和订单表的关系模型

06

第6章　MyBatis入门

学习目标

- 了解 MyBatis 的优缺点
- 了解 MyBatis 与 JDBC 和 Hibernate 的区别
- 掌握 MyBatis 的工作流程
- 熟悉 MyBatis 的入门案例
- 能够使用分层思想整理及优化 MyBatis 的入门案例

6.1　MyBatis 概述

MyBatis 概述

MyBatis 的原名叫 iBatis，来源于“internet”和“abatis”的组合，是一个 2001 年发起的开放源代码项目。MyBatis 最初侧重于密码软件的开发，现在则是一个基于 Java 的持久层框架。

MyBatis 让程序员将主要精力放在 SQL 上，通过 MyBatis 提供的映射方式，自由灵活地生成（半自动化，大部分需要程序员编写 SQL）满足需求的 SQL 语句。MyBatis 可以将 preparedStatement 中的输入参数自动进行输入映射，将查询结果集灵活映射成 Java 对象。

MyBatis 框架的优点如下。

① 与 JDBC 相比，减少了 50%以上的代码量。

② MyBatis 是最简单的持久化框架，小巧且简单易学。

③ MyBatis 灵活，不会对应用程序或者数据库的现有设计强加任何影响，SQL 代码写在 XML 文件里，实现了从程序代码（Java 文件）中彻底分离，降低了耦合度，便于统一管理、优化和可重用。

④ 提供 XML 标签，支持编写动态 SQL 语句（XML 中使用 if、else）。

⑤ 提供映射标签，支持对象与数据库的 ORM 字段关系映射（在 XML 中配置映射关系，也可以使用注解）。

MyBatis 框架的缺点如下。

① SQL 语句的编写工作量较大，尤其是字段多、关联表多时，更

是如此，对开发人员编写 SQL 语句的功底有一定要求。

② SQL 语句依赖于数据库，导致数据库移植性差，不能随意更换数据库。

MyBatis 专注于 SQL 本身，是一个足够灵活的 DAO 层解决方案。因此，MyBatis 框架适用于对性能的要求很高或者需求变化较多的项目，如互联网项目。

下面简述 MyBatis 与 JDBC 和 Hibernate 的区别。

（1）MyBatis 与传统的 JDBC 进行比较

MyBatis 很简单，对 JDBC 的封装很好，比 JDBC 减少了 62%的代码量；MyBatis 把 SQL 代码从程序代码中彻底分离，使 SQL 代码可重用；MyBatis 增强了项目中的分工；MyBatis 增强了代码的移植性。

（2）MyBatis 与 Hibernate 进行比较

MyBatis 与 Hibernate 框架的映射关系不同。

Hibernate 对数据库结构提供了较为完整的封装，Hibernate 的 ORM 实现了 POJO 和数据库表之间的映射，以及 SQL 的自动生成和执行，只要定义好了 POJO 到数据库表的映射关系，即能通过提供的方法完成持久层操作。Hibernate 不需要程序员熟练掌握 SQL，因为它会根据制定的存储逻辑，自动生成对应的 SQL 并调用 JDBC 接口加以执行。Hibernate 的映射关系如图 6-1 所示。

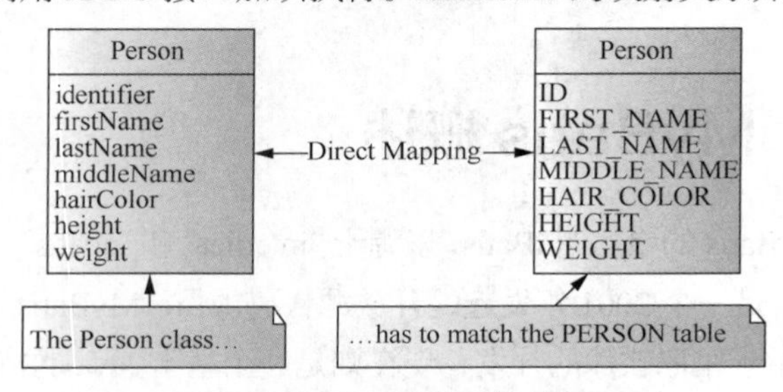

图 6-1　Hibernate 的映射关系

MyBatis 框架是 POJO 与 SQL 之间的映射关系。通过映射配置文件，将 SQL 所需的参数和返回的结果字段映射到指定 POJO。相对“O/R”而言，MyBatis 是一种“SQL Mapping（SQL 映射）”的 ORM 实现。MyBatis 的映射关系如图 6-2 所示。

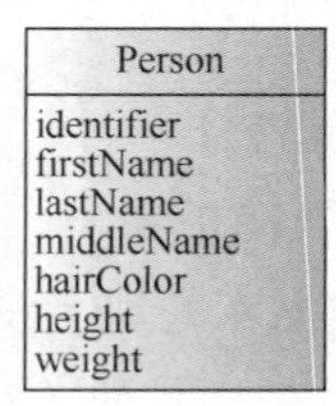


```
SELECT
  ID            as identifier,
  FIRST_NAME    as firstName,
  LAST_NAME     as lastName,
  MIDDLE_NAME   as middleName,
  HAIR_COLOR    as hairColor,
  HEIGHT        as height,
  WEIGHT        as weight
FROM PERSON
WHERE ID = #identifier#
```

Person
ID
FIRST_NAME
LAST_NAME
MIDDLE_NAME
HAIR_COLOR
HEIGHT
WEIGHT

图 6-2　MyBatis 的映射关系

6.2　MyBatis 的工作流程

MyBatis 的工作流程

MyBatis 的工作流程如图 6-3 所示。

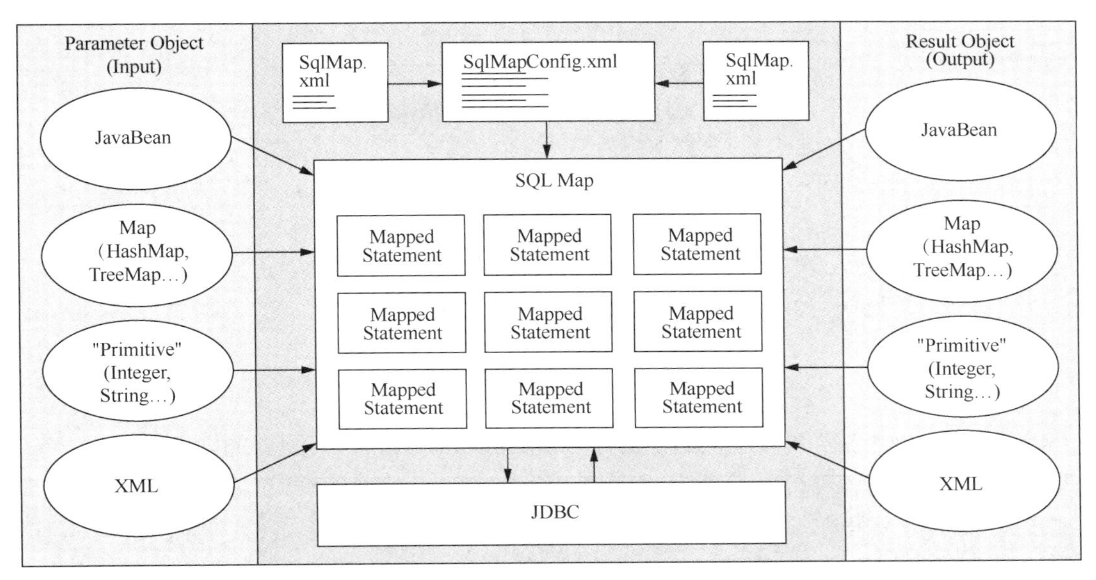

图 6-3　MyBatis 的工作流程

从图 6-3 中可以看到，MyBatis 的工作流程包括如下 4 步。

（1）第 1 步，加载配置文件（mybatis-config.xml、*...Mapper.xml）并初始化，将 SQL 的配置信息加载成为一个个 MappedStatement 对象（包括了传入参数映射配置、执行的 SQL 语句、结果映射配置），并存储在内存中。

（2）第 2 步，接收调用请求（调用 MyBatis 提供的 API，即增删改查的方法）并传入参数，即 SQL 的 ID 和传入参数对象。

（3）第 3 步，处理操作请求，具体过程如下。

① 根据 SQL 的 ID 查找对应的 MappedStatement 对象。

② 根据传入参数对象解析 MappedStatement 对象，得到最终要执行的 SQL 和传入参数。

③ 获取数据库连接，根据得到的最终 SQL 语句和传入参数，到数据库执行，并得到执行结果。

④ 根据 MappedStatement 对象中的结果映射配置对得到的执行结果进行转换处理，并得到最终的处理结果。

⑤ 释放连接资源。

（4）第 4 步，返回处理结果。

6.3　MyBatis 的入门案例

MyBatis 的入门案例

使用 MyBatis 的配置步骤如下。

（1）在数据库中创建测试表。

（2）在 Eclipse 集成开发工具中新建一个工程。

（3）在工程中引入相关的 JAR 包。使用 MyBatis 框架需要引入 MyBatis 的包，如 mybatis-3.4.4.jar、ojdbc6.jar（数据库的驱动包）等，如图 6-4 所示。

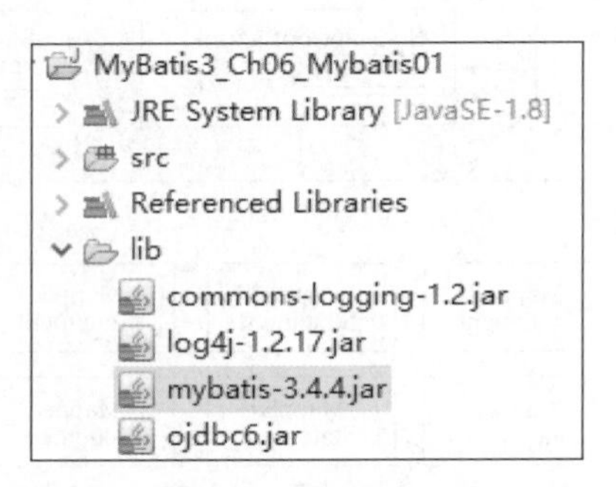

图 6-4　使用 MyBatis 需引入的 JAR 包

（4）在工程的类路径下创建总配置文件 SqlMapConfig.xml，其配置的内容如下。

```
<configuration>
    <!-- 实体类的别名 -->
    <typeAliases>
        <typeAlias type="cn.MyBatis.domain.Student" alias="Student"/>
    </typeAliases>
    <!-- 环境管理，可以配置多个环境，默认是 default="development"的环境 -->
    <environments default="development">
        <environment id="development">
            <!-- 事务管理方式，JDBC 事务 -->
            <transactionManager type="JDBC" />
            <!-- 数据源，type="POOLED"表示使用连接池 -->
            <dataSource type="POOLED">
                <property value="oracle.jdbc.OracleDriver" name="driver" />
                <property value="jdbc:oracle:thin:@localhost:1521/xe" name="url" />
                <property value="scott" name="username" />
                <property value="tiger" name="password" />
            </dataSource>
        </environment>
    </environments>
    <!--映射文件路径，可能会有多个 mapper 映射 -->
    <mappers>
       <mapper resource="cn/mybatis/dao/studentMapper.xml"/>
    </mappers>
</configuration>
```

<property>的配置提供了允许在主配置文件之外的一个“名值对”列表，可以将其中的配置信息加载进来，而这些配置信息可以放在任何一个地方。使用 properties 元素，其中有两个属性，分别是 resource 和 url。MyBatis 将按照下面的顺序来加载属性：在 properties 元素体内定义的属性首先被读取，然后会读取 properties 元素中 resource 或 url 加载的属性，它们会覆盖已读取的同名属性。

<mapper>指定了映射文件的位置，配置文件中<mappers>下可以定义多个<mapper>，以指定项目内包含的所有映射文件。

（5）在类路径下创建 log4j.properties。

下面看一个 MyBatis 的入门案例。

【实例 6.1】 MyBatis 入门案例。

通过向学生表中插入一条学生记录，讲解 MyBatis 的使用方法（代码详见 MyBatis3_Ch06_Mybatis01\src\cn\mybatis\domain\Student.java；MyBatis3_Ch06_Mybatis01\src\cn\mybatis\dao\studentMapper.xml；MyBatis3_Ch06_Mybatis01\src\cn\mybatis\demo\Demo_01.java）。

（1）在数据库中创建测试表 Student，如表 6-1 所示。

表 6-1　Student 表

Field	Type
stu_id	int(11) NOT NULL
stu_name	varchar(50) NULL
stu_birthdate	date NULL
stu_phone	varchar(50) NULL

（2）在已建好并引入 MyBatis 相关 JAR 包的工程中，创建 pojo 类。

```
package cn.mybatis.domain;
import java.util.Date;
public class Student {
    private int stuId;
    private String stuName;
    private Date stuBirthdate;
    private String stuPhone;
    public int getStuId() {
        return stuId;
    }
    public void setStuId(int stuId) {
        this.stuId = stuId;
    }
    public String getStuName() {
        return stuName;
    }
    public void setStuName(String stuName) {
        this.stuName = stuName;
    }
    public Date getStuBirthdate() {
        return stuBirthdate;
    }
    public void setStuBirthdate(Date stuBirthdate) {
        this.stuBirthdate = stuBirthdate;
    }
    public String getStuPhone() {
        return stuPhone;
    }
    public void setStuPhone(String stuPhone) {
        this.stuPhone = stuPhone;
    }
}
```

（3）创建配置 Mapper 映射文件。在 Mapper 映射文件中配置 ResultMap 类表映射和使用<insert><update><delete><select>等标签写入 SQL 语句。

```
<?xml version="1.0" encoding="UTF-8" ?>
<!DOCTYPE mapper PUBLIC "-//mybatis.org//DTD Mapper 3.0//EN"
"http://mybatis.org/dtd/mybatis-3-mapper.dtd">
<mapper namespace="Student">
    <resultMap type="cn.mybatis.domain.Student" id="studentResultMap">
        <id column="stu_id" property="stuId" jdbcType="INTEGER"
            javaType="java.lang.Integer" />
        <result column="stu_name" property="stuName" jdbcType="VARCHAR"
            javaType="java.lang.String" />
        <result column="stu_birthdate" property="stuBirthdate"
            jdbcType="DATE" javaType="java.util.Date" />
        <result column="stu_phone" property="stuPhone" jdbcType="VARCHAR"
            javaType="java.lang.String" />
    </resultMap>
    <insert id="insertStudent" parameterType="cn.mybatis.domain.Student">
        insert into student (stu_id,stu_name,stu_birthdate,stu_phone)
    values(#{stuId},#{stuName},#{stuBirthdate},#{stuPhone})
    </insert>
</mapper>
```

（4）在核心配置文件 SqlMapConfig.xml 中，用<mapper/>标签引入映射文件路径。

```
<mappers>
        <mapper resource="cn/mybatis/dao/studentMapper.xml" />
</mappers>
```

（5）编写程序 Demo_01 读取核心配置文件。

```
package cn.mybatis.demo;
import java.io.InputStream;
import java.text.SimpleDateFormat;
import org.apache.ibatis.io.Resources;
import org.apache.ibatis.session.SqlSession;
import org.apache.ibatis.session.SqlSessionFactory;
import org.apache.ibatis.session.SqlSessionFactoryBuilder;
import org.apache.log4j.Logger;
import cn.mybatis.domain.Student;
public class Demo_01 {
    public static void main(String[] args) throws Exception {
        //获取配置信息
        InputStream is = Resources.getResourceAsStream("SqlMapperConfig.xml");
        //获取 session 工厂类
        SqlSessionFactory fac=new SqlSessionFactoryBuilder().build(is);
        //创建要保存的学生信息
        Student s=new Student();
        s.setStuId(1);
        s.setStuName("tom");
        s.setStuBirthdate(new SimpleDateFormat("yyyy-MM-dd").parse("1990-11-3"));
        s.setStuPhone("1505315****");
        SqlSession session = null;
        try {
            //调用工厂类的 openSession 方法获取数据库的 session
            session = fac.openSession();
            /**
             *调用 session 的 insert 方法实现保存功能
```

```
                *其参数有两个
                 *1.Student.insertStudent 表示 Student 命名空间下的
                 *id 为 insertStudent 的 SQL
                *2.要保存的 SQL
                 */
            session.insert("Student.insertStudent", s);
                session.commit();
            } catch (Exception e) {
                e.printStackTrace();
            } finally {
                session.close();
            }
        }
    }
```

选中 Demo_01 文件，单击鼠标右键，选择“Run As”菜单中的子菜单“1 Java Application”，运行后，数据库的 Student 表中添加了一条记录，如图 6-5 所示。

	STU_ID	STU_NAME	STU_BIRTHDATE	STU_PHONE
1	1	tom ...	1990/11/3	1505315**** ...

图 6-5 在 Student 表中插入的数据

【程序解析】

在 Mapper 映射文件中进行如下配置。

① ResultMap 类表映射。

② <insert><update><delete><select>标签写入 SQL 语句。

```
<mapper namespace="student">
```

具体配置如下。

```
</mapper>
```

以上代码中，mapper 是根元素，namespace 属性配置命名空间，命名空间可以用来有效地区分 SQL 语句的 id。

```
<resultMap id="BaseResultMap" type="com.icss.MyBatis.pojo.Student">
<id column="stu_id" property="stuId" jdbcType="INTEGER" />
<result column="stu_name" property="stuName" jdbcType="VARCHAR" />
<result column="stu_birthdate" property="stuBirthdate" jdbcType="TIMESTAMP" />
<result column="stu_phone" property="stuPhone" jdbcType="VARCHAR" />
</resultMap>
```

以上代码中，<resultMap>配置表列与类属性映射的结果集，type 属性可以使用完整 pojo 类名，如果定义了类别名，也可以使用类别名。不需要把所有字段都映射，可以只映射一部分，主键字段必须用 id 元素指定。

```
<insert id="insert" parameterType="com.icss.MyBatis.pojo.Student">
insert into student(stu_id, stu_name, stu_birthdate, stu_phone)
values(default, #{stuName},#{stuBirthdate},#{stuPhone} )
</insert>
```

以上代码插入数据，#{}语法用于插入动态数据，数据来源于调用此语句时传入的参数，参数类型由 parameterType 属性决定。

软件分层的好处有很多：软件分层可以使软件具有结构性，便于用户开发、维护和管理；软件分层可以将不同功能的模块独立起来，以便进行代码的复用和替换。下面通过实例演示利用分层思想整理和优化 MyBatis 的入门案例。

【实例 6.2】使用分层思想整理和优化 MyBatis 的入门案例演示。

相关代码详见 MyBatis3_Ch06_Mybatis01\cn\mybatis\dao\studentMapper.xml；MyBatis3_Ch06_Mybatis01\cn\mybatis\dao\StudentDao.java；MyBatis3_Ch06_Mybatis01\cn\mybatis\dao\impl\StudentDaoImpl.java；MyBatis3_Ch06_Mybatis01\cn\mybatis\domain\Student.java；MyBatis3_Ch06_Mybatis01\cn\mybatis\demo\Demo_02.java。

（1）studentMapper.xml 和 Student.java

其内容与实例 6.1 中的相同。

（2）StudentDao.java

```
package cn.mybatis.dao;
import cn.mybatis.domain.Student;
public interface StudentDao {
    public void insertStudent(Student s);
}
```

（3）StudentDaoImpl.java

```
package cn.mybatis.dao.impl;
import org.apache.ibatis.session.SqlSession;
import org.apache.ibatis.session.SqlSessionFactory;
import cn.mybatis.dao.StudentDao;
import cn.mybatis.domain.Student;
public class StudentDaoImpl implements StudentDao {
    private SqlSessionFactory fac;
    public StudentDaoImpl(SqlSessionFactory fac) {
        this.fac = fac;
    }
    @Override
    public void insertStudent(Student s) {
        SqlSession sess = null;
        try {
            sess = this.fac.openSession();
            sess.insert("Student.insertStudent", s);
            sess.commit();
        } catch (Exception e) {
            e.printStackTrace();
        } finally {
            sess.close();
        }
    }
}
```

（4）Demo_02.java

```
package cn.mybatis.demo;
```

```
import java.io.InputStream;
import java.text.SimpleDateFormat;
import org.apache.ibatis.io.Resources;
import org.apache.ibatis.session.SqlSessionFactory;
import org.apache.ibatis.session.SqlSessionFactoryBuilder;
import org.apache.log4j.Logger;
import cn.mybatis.dao.StudentDao;
import cn.mybatis.dao.impl.StudentDaoImpl;
import cn.mybatis.domain.Student;
public class Demo_02 {
    private static SqlSessionFactory fac;
    static{
        InputStream is=null;
        try {
            //获取配置信息
            is = Resources.getResourceAsStream("SqlMapperConfig.xml");
            //获取 session 工厂类
            fac=new SqlSessionFactoryBuilder().build(is);
        } catch (Exception e) {
            e.printStackTrace();
            Logger.getLogger(Demo_02.class).debug(e.getMessage());
        }
    }
    public static void main(String[] args) throws Exception {
        Student s=new Student();
        s.setStuId(1);
        s.setStuName("tom");
        s.setStuBirthdate(new SimpleDateFormat("yyyy-MM-dd").parse("2001-12-3"));
        s.setStuPhone("1505315****");
        StudentDao dao=new StudentDaoImpl(fac);
        dao.insertStudent(s);
    }
}
```

选中 Demo_02 文件，单击鼠标右键，选择“Run As”菜单中的子菜单“1 Java Application”，运行后，数据库的 Student 表中添加了一条记录，如图 6-6 所示。

	STU_ID	STU_NAME	STU_BIRTHDATE	STU_PHONE
1	1	tom ···	2001/12/3 ▾	1505315**** ···

图 6-6 在 Student 表中插入的数据

本章小结

本章对 MyBatis 的概念与优缺点、MyBatis 与 JDBC 和 Hibernate 的比较、MyBatis 的工作流程和环境配置进行了较为详细的讲解，此外还讲解了 MyBatis 的入门案例以及使用分层思想整理和优化 MyBatis 的入门案例。通过本章的学习，读者可以理解 MyBatis 的概念和工作流程，能够进行 MyBatis 的环境配置，并开发一个 MyBatis 的案例。

习题

1. 简述 MyBatis 的核心配置文件 sqlmapconfig.xml 的内容。
2. 简述 MyBatis 框架相较 JDBC 的优点。

上机指导

1. 使用原生 JDBC 编写操作 Oracle 数据库的程序（查询 person 表中 name=小明的用户信息）。

2. 使用 MyBatis 框架编写操作 Oracle 数据库的程序（根据 id=12 查询 tb_user 表中的用户信息）。

07 第7章　基于MyBatis的增删改查操作

学习目标

- 掌握基于 MyBatis 的增删改查操作方法
- 掌握 MyBatis 的结果类型（resultMap 和 resultType）的用法
- 掌握 MyBatis 的参数类型的用法
- 了解#和$的用法

本章将通过使用 MyBatis 对 Student 表进行增删改查的案例，讲解 MyBatis 框架的基本应用。

7.1　基于 MyBatis 的数据添加

基于 MyBatis 的数据添加

基于 MyBatis 的数据添加，可参见第 6 章的实例 6.1。我们再把例子优化一下。

【实例 7.1】基于 MyBatis 的数据添加案例。

通过 MyBatis 往 Oracle 数据库的 Student 表中插入数据。

准备工作：Oracle 数据库中建有 scott 账号，密码为 tiger。scott 账号下有 Student 表，表中有字段学生 Id（STU_ID）、学生名称（STU_NAME）、学生出生日期（STU_BIRTHDATE）、学生电话（STU_PHONE）。

在 scott 账号下创建 SEQSTU 序列，以实现 Student 表主键学生 Id 的自增功能。创建序列的语句如下所示。

```
CREATE SEQUENCE SEQSTU
MINVALUE 1
START WITH 1
INCREMENT BY 1
NOCACHE
NOCYCLE;
```

创建一个 Web 工程，并引入 MyBatis 相关 JAR 包和数据库驱动包。

具体源代码如下所示（代码详见 MyBatis3_Ch07_Mybatis02\src\cn\mybatis\dao\studentMapper.xml；MyBatis3_Ch07_Mybatis02\src\cn\mybatis\dao\StudentDao.java；MyBatis3_Ch07_Mybatis02\src\cn\mybatis\dao\impl\

StudentDaoImpl.java；MyBatis3_Ch07_Mybatis02\src\cn\mybatis\domain\Student.java；MyBatis3_Ch07_Mybatis02\src\cn\mybatis\demo\Demo_01.java）。

（1）studentMapper.xml

```
<?xml version="1.0" encoding="UTF-8" ?>
<!DOCTYPE mapper PUBLIC "-//mybatis.org//DTD Mapper 3.0//EN"
"http://mybatis.org/dtd/mybatis-3-mapper.dtd">
<mapper namespace="Student">
   <resultMap type="cn.mybatis.domain.Student" id="studentResultMap">
       <id column="stu_id" property="stuId" jdbcType="INTEGER" javaType=
           "java.lang.Integer" />
       <result column="stu_name" property="stuName" jdbcType="VARCHAR"
           javaType="java.lang.String" />
       <result column="stu_birthdate" property="stuBirthdate"
           jdbcType="DATE" javaType="java.util.Date" />
       <result column="stu_phone" property="stuPhone" jdbcType="VARCHAR"
           javaType="java.lang.String" />
    </resultMap>
   <insert id="insertStudent" parameterType="cn.mybatis.domain.Student">
       insert into student(stu_id,stu_name,stu_birthdate,stu_phone)
   values(seqstu.nextval,#{stuName},#{stuBirthdate},#{stuPhone})
   </insert>
</mapper>
```

（2）StudentDao.java

```
package cn.mybatis.dao;
import java.util.List;
import cn.mybatis.domain.Student;
public interface StudentDao {
    public void insertStudent(Student s);
}
```

（3）StudentDaoImpl.java

```
package cn.mybatis.dao.impl;
import java.sql.Date;
import java.util.HashMap;
import java.util.List;
import java.util.Map;
import org.apache.ibatis.session.SqlSession;
import org.apache.ibatis.session.SqlSessionFactory;
import cn.mybatis.dao.StudentDao;
import cn.mybatis.domain.Student;
public class StudentDaoImpl implements StudentDao {
    private SqlSessionFactory fac;
    public StudentDaoImpl(SqlSessionFactory fac) {
        this.fac = fac;
    }
    @Override
    public void insertStudent(Student s) {
        SqlSession sess = null;
        try {
            sess = this.fac.openSession();
            sess.insert("Student.insertStudent", s);
            sess.commit();
```

```
        } catch (Exception e) {
            e.printStackTrace();
        } finally {
            sess.close();
        }
    }
```

（4）Student.java

```
package cn.mybatis.domain;
import java.util.Date;
public class Student {
    private int stuId;
    private String stuName;
    private Date stuBirthdate;
    private String stuPhone;
    public int getStuId() {
        return stuId;
    }
    public void setStuId(int stuId) {
        this.stuId = stuId;
    }
    public String getStuName() {
        return stuName;
    }
    public void setStuName(String stuName) {
        this.stuName = stuName;
    }
    public Date getStuBirthdate() {
        return stuBirthdate;
    }
    public void setStuBirthdate(Date stuBirthdate) {
        this.stuBirthdate = stuBirthdate;
    }
    public String getStuPhone() {
        return stuPhone;
    }
    public void setStuPhone(String stuPhone) {
        this.stuPhone = stuPhone;
    }
    @Override
    public String toString() {
        return "Student [stuId=" + stuId + ", stuName=" + stuName + ",
        stuBirthdate=" + stuBirthdate + ", stuPhone=" + stuPhone + "]";
    }
}
```

（5）Demo_01.java

```
import org.apache.ibatis.io.Resources;
import org.apache.ibatis.session.SqlSessionFactory;
import org.apache.ibatis.session.SqlSessionFactoryBuilder;
import org.apache.log4j.Logger;
import org.junit.Test;
import cn.mybatis.dao.StudentDao;
import cn.mybatis.dao.impl.StudentDaoImpl;
import cn.mybatis.domain.Student;
```

```
public class Demo_01 {
    private static SqlSessionFactory fac;
    static{
        InputStream is=null;
        try {
            //处理并根据 config 配置文件实例化 SqlSessionFactory
            is = Resources.getResourceAsStream("SqlMapperConfig.xml");
            fac=new SqlSessionFactoryBuilder().build(is);
        } catch (Exception e) {
            e.printStackTrace();
            Logger.getLogger(Demo_01.class).debug(e.getMessage());
        }
    }
    public static void main(String[] args) throws Exception {
        Student s=new Student();
        s.setStuName("tom");
        s.setStuBirthdate(new SimpleDateFormat("yyyy-MM-dd").parse("2001-12-3"));
        s.setStuPhone("1505315****");
        StudentDao dao=new StudentDaoImpl(fac);
        dao.insertStudent(s);
    }
```

运行程序。选中 Demo_01.java 文件，单击鼠标右键，选择“Run As”菜单中的子菜单“1 JavaApplication”，运行程序后数据库添加了新记录，操作结果如图 7-1 所示。

	STU_ID	STU_NAME	STU_BIRTHDATE	STU_PHONE
1	1	tom	2001/12/3	1505315****

图 7-1　基于 MyBatis 的添加操作结果

因为 Student 表中主键是自增的，所以可以多次运行添加程序，每运行一次添加程序，数据库中就会添加一条记录。如再运行两次添加程序，Student 表中添加的数据如图 7-2 所示。

	STU_ID	STU_NAME	STU_BIRTHDATE	STU_PHONE
1	1	tom	2001/12/3	1505315****
2	2	tom	2001/12/3	1505315****
3	3	tom	2001/12/3	1505315****

图 7-2　基于 MyBatis 的添加再次操作结果

7.2　基于 MyBatis 的数据删除

基于 MyBatis 的数据删除

基于 MyBatis 的数据删除操作在映射文件中是通过配置 delete 元素来实现的。操作思路和编码过程具体见下面的实例。

【实例 7.2】基于 MyBatis 的数据删除案例。

通过 MyBatis 在 Oracle 数据库的 Student 表中删除一条记录。

准备工作：参见实例 7.1 的准备工作。

具体源代码如下所示（代码详见 MyBatis3_Ch07_Mybatis02\src\cn\mybatis\dao\studentMapper.xml；MyBatis3_Ch07_Mybatis02\src\cn\mybatis\dao\StudentDao.java；MyBatis3_Ch07_Mybatis02\src\cn\mybatis\dao\impl\StudentDaoImpl.java；MyBatis3_Ch07_Mybatis02\src\cn\mybatis\demo\Demo_01.java）。

（1）studentMapper.xml

```
<?xml version="1.0" encoding="UTF-8" ?>
<!DOCTYPE mapper PUBLIC "-//mybatis.org//DTD Mapper 3.0//EN"
"http://mybatis.org/dtd/mybatis-3-mapper.dtd">
<mapper namespace="Student">
   <delete id="deleteStudent" parameterType="java.lang.Integer">
       delete from student
       where stu_id=#{stuId}
   </delete>
</mapper>
```

（2）StudentDao.java

在实例 7.1 的 StudentDao.java 基础上添加以下代码。

```
public void deleteStudent(int stuId);
```

（3）StudentDaoImpl.java

```
@Override
public void deleteStudent(int stuId) {
    SqlSession sess = null;
    try {
        sess = this.fac.openSession();
        sess.delete("Student.deleteStudent", stuId);
        sess.commit();
    } catch (Exception e) {
        e.printStackTrace();
    } finally {
        sess.close();
    }
}
```

（4）Demo_01.java

```
@Test
public void testDelete(){
    StudentDao dao=new StudentDaoImpl(fac);
    dao.deleteStudent(1);
}
```

运行程序。选中 Package Explorer 视图下 Demo_01.java 文件下的方法 testDelete，单击鼠标右键，选择“Run As”菜单中的子菜单“2 Junit Test”，运行程序后数据库删除了那条记录，操作结果如图 7-3 所示。

	STU_ID	STU_NAME	STU_BIRTHDATE	STU_PHONE
▶ 1	2	tom ...	2001/12/3 ▾	1505315**** ...
2	3	tom ...	2001/12/3 ▾	1505315**** ...

图 7-3　基于 MyBatis 的删除操作结果

7.3　基于 MyBatis 的数据修改

MyBatis 的更新操作在映射文件中是通过配置 update 元素来实现的。基于 MyBatis 的数据修改的操作思路和编码过程具体见下面的实例。

基于 MyBatis 的数据修改

【实例 7.3】基于 MyBatis 的数据修改案例。

通过 MyBatis 在 Oracle 数据库的 Student 表中修改一条记录。

准备工作：参见实例 7.1 的准备工作。

具体源代码如下所示（代码详见 MyBatis3_Ch07_Mybatis02\src\cn\mybatis\dao\studentMapper.xml；MyBatis3_Ch07_Mybatis02\src\cn\mybatis\dao\StudentDao.java；MyBatis3_Ch07_Mybatis02\src\cn\mybatis\dao\impl\StudentDaoImpl.java；MyBatis3_Ch07_Mybatis02\src\cn\mybatis\domain\Student.java；MyBatis3_Ch07_Mybatis02\src\cn\mybatis\demo\Demo_01.java）。

（1）studentMapper.xml

```
<?xml version="1.0" encoding="UTF-8" ?>
<!DOCTYPE mapper PUBLIC "-//mybatis.org//DTD Mapper 3.0//EN"
"http://mybatis.org/dtd/mybatis-3-mapper.dtd">
<mapper namespace="Student">
    <update id="updateStudent" parameterType="cn.mybatis.domain.Student">
        update student set
    stu_name=#{stuName},stu_birthdate=#{stuBirthdate},stu_phone=#{stuPhone}
    where stu_id=#{stuId}
    </update>
</mapper>
```

（2）StudentDao.java

在实例 7.1 的 StudentDao.java 基础上添加以下代码。

```
public void updateStudent(Student s);
```

（3）StudentDaoImpl.java

```
@Override
public void updateStudent(Student s) {
    SqlSession session = null;
    try {
        session = this.fac.openSession();
        session.update("Student.updateStudent", s);
        session.commit();
    } catch (Exception e) {
        e.printStackTrace();
    } finally {
        session.close();
    }
}
```

（4）Student.java

与实例 7.1 的 Student.java 相同。

（5）Demo_01.java

```
@Test
public void testUpdate() throws Exception{
    Student s=new Student();
    s.setStuId(1);
    s.setStuName("tom2");
    s.setStuBirthdate(new SimpleDateFormat("yyyy-MM-dd").parse("2000-12-3"));
    s.setStuPhone("1505318****");
    StudentDao dao=new StudentDaoImpl(fac);
```

```
        dao.updateStudent(s);
    }
```

运行程序。选中 Package Explorer 视图下 Demo_01.java 文件下的方法 testUpdate，单击鼠标右键，选择“Run As”菜单中的子菜单“2 Junit Test”，运行程序后数据库修改了那条记录，操作结果如图 7-4 所示。

	STU_ID	STU_NAME	STU_BIRTHDATE	STU_PHONE
1	1	tom2	2000/12/3	1505318****
2	2	tom	2001/12/3	1505315****
3	3	tom	2001/12/3	1505315****

图 7-4　基于 MyBatis 的修改操作结果

【案例分析】

与在 Student 表中添加数据的方法相比，修改操作的代码增加了 STU_ID 属性值的设置，并调用 SqlSession 的 update 方法对 STU_ID 为 1 的学生的姓名、生日和电话进行修改。

7.4　基于 MyBatis 的数据查询

基于 MyBatis 的数据查询

基于 MyBatis 的数据查询包括查询表中的单条记录、查询表中的多条记录等，下面进行详细介绍。

7.4.1　单条记录的查询

基于 MyBatis 的单条记录查询的操作思路和编码过程具体见下面的实例。

【实例 7.4】基于 MyBatis 的单条记录查询案例。

通过 MyBatis 查询 Student 表中的一条记录。

准备工作：见实例 7.1 的准备工作。

具体源代码如下所示（代码详见 MyBatis3_Ch07_Mybatis02\src\cn\mybatis\dao\studentMapper.xml；MyBatis3_Ch07_Mybatis02\src\cn\mybatis\dao\StudentDao.java；MyBatis3_Ch07_Mybatis02\src\cn\mybatis\dao\impl\StudentDaoImpl.java；MyBatis3_Ch07_Mybatis02\src\cn\mybatis\domain\Student.java；MyBatis3_Ch07_Mybatis02\src\cn\mybatis\demo\Demo_01.java）。

（1）studentMapper.xml

```
<?xml version="1.0" encoding="UTF-8" ?>
<!DOCTYPE mapper PUBLIC "-//mybatis.org//DTD Mapper 3.0//EN"
"http://mybatis.org/dtd/mybatis-3-mapper.dtd">
<mapper namespace="Student">
   <resultMap type="cn.mybatis.domain.Student" id="studentResultMap">
      <id column="stu_id" property="stuId" jdbcType="INTEGER"
         javaType="java.lang.Integer" />
      <result column="stu_name" property="stuName" jdbcType="VARCHAR"
         javaType="java.lang.String" />
      <result column="stu_birthdate" property="stuBirthdate"
         jdbcType="DATE" javaType="java.util.Date" />
      <result column="stu_phone" property="stuPhone"
         jdbcType="VARCHAR" javaType="java.lang.String" />
```

```
    </resultMap>
    <select id="selectStudentById" parameterType="java.lang.Integer"
        resultMap="studentResultMap">
        select * from student where stu_id=#{stuId}
    </select>
</mapper>
```

（2）StudentDao.java

在实例 7.1 的 StudentDao.java 基础上添加以下代码。

```
public Student selectStudentById(int stuId);
```

（3）StudentDaoImpl.java

```
@Override
public Student selectStudentById(int stuId) {
    SqlSession sess = null;
    try {
        sess = this.fac.openSession();
        Student s = sess.selectOne("Student.selectStudentById", stuId);
        sess.commit();
        return s;
    } catch (Exception e) {
        e.printStackTrace();
    } finally {
        sess.close();
    }
    return null;
}
```

（4）Student.java

与实例 7.1 中的 Student.java 相同。

（5）Demo_01.java

```
@Test
public void selectByIdTest(){
    StudentDao dao=new StudentDaoImpl(fac);
    Student student = dao.selectStudentById(2);
    System.out.println(student.getStuName()+"  "+student.getStuBirthdate()+"  "+
    student.getStuPhone());
}
```

运行程序。选中 Package Explorer 视图下 Demo_01.java 文件下的方法 selectByIdTest，单击鼠标右键，选择“Run As”菜单中的子菜单“2 Junit Test”，运行程序后控制台显示如图 7-5 所示。

```
tom  Mon Dec 03 00:00:00 CST 2001  1505315****
```

图 7-5　基于 MyBatis 的单条记录查询操作结果

7.4.2　多条记录的查询

基于 MyBatis 的多条记录查询的操作思路和编码过程具体见下面的实例。

【实例 7.5】基于 MyBatis 的多条记录查询案例。

通过MyBatis查询Student表中的所有记录。

准备工作：见实例7.1的准备工作。

具体源代码如下所示（代码详见MyBatis3_Ch07_Mybatis02\src\cn\mybatis\dao\studentMapper.xml；MyBatis3_Ch07_Mybatis02\src\cn\mybatis\dao\StudentDao.java；MyBatis3_Ch07_Mybatis02\src\cn\mybatis\dao\impl\StudentDaoImpl.java；MyBatis3_Ch07_Mybatis02\src\cn\mybatis\domain\Student.java；MyBatis3_Ch07_Mybatis02\src\cn\mybatis\demo\Demo_01.java）。

（1）studentMapper.xml

```
<?xml version="1.0" encoding="UTF-8" ?>
<!DOCTYPE mapper PUBLIC "-//mybatis.org//DTD Mapper 3.0//EN"
"http://mybatis.org/dtd/mybatis-3-mapper.dtd">
<mapper namespace="Student">
   <resultMap type="cn.mybatis.domain.Student" id="studentResultMap">
       <id column="stu_id" property="stuId" jdbcType="INTEGER"
           javaType="java.lang.Integer" />
       <result column="stu_name" property="stuName" jdbcType="VARCHAR"
           javaType="java.lang.String" />
       <result column="stu_birthdate" property="stuBirthdate"
           jdbcType="DATE" javaType="java.util.Date" />
       <result column="stu_phone" property="stuPhone" jdbcType="VARCHAR"
           javaType="java.lang.String" />
   </resultMap>
   <select id="selectStudent" resultMap="studentResultMap">
       select * from student
   </select>
</mapper>
```

（2）StudentDao.java

```
public List<Student> selectStudent();
```

（3）StudentDaoImpl.java

```
@Override
public List<Student> selectStudent() {
    SqlSession sess = null;
    try {
        sess = this.fac.openSession();
        List<Student> list = sess.selectList("Student.selectStudent");
        sess.commit();
        return list;
    } catch (Exception e) {
        e.printStackTrace();
    } finally {
        sess.close();
    }
    return null;
}
```

（4）Student.java

与实例7.1中的Student.java相同。

（5）Demo_01.java

```
@Test
public void selectStudentTest(){
    StudentDao dao=new StudentDaoImpl(fac);
    List<Student> list=dao.selectStudent();
    System.out.println(list);
}
```

运行程序。选中 Package Explorer 视图下 Demo_01.java 文件下的方法 selectStudentTest，单击鼠标右键，选择“Run As”菜单中的子菜单“2 Junit Test”，运行程序后控制台显示如图 7-6 所示。

```
[Student [stuId=2, stuName=tom, stuBirthdate=Mon Dec 03 00:00:00 CST 2001, stuPhone=1505315****],
Student [stuId=3, stuName=tom, stuBirthdate=Mon Dec 03 00:00:00 CST 2001, stuPhone=1505315****]]
```

图 7-6　基于 MyBatis 的多条记录查询操作结果

7.5　MyBatis 的结果类型

MyBatis 的结果类型

MyBatis 的结果类型有 resultMap 和 resultType 两种，这两种结果类型同时使用是无意义的，必须分开使用。

7.5.1　resultMap

属性值是已经定义好的<resultMap>的 id。

示例如下：

```
<resultMap type="cn.mybatis.domain.Student" id="studentResultMap">
    <id column="stu_id" property="stuId" jdbcType="INTEGER"
        javaType="java.lang.Integer" />
    <result column="stu_name" property="stuName" jdbcType="VARCHAR"
        javaType="java.lang.String" />
    <result column="stu_birthdate" property="stuBirthdate"
        jdbcType="DATE" javaType="java.util.Date" />
    <result column="stu_phone" property="stuPhone" jdbcType="VARCHAR"
        javaType="java.lang.String" />
    </resultMap>
<select id="query" resultMap="BaseResultMap">
```

在上面的例子中，resultMap 元素定义了一个 id“studentResultMap”，它的属性 id 代表它的标识，type 代表使用哪种类作为其映射的类。子元素 id 代表 resultMap 的主键，而 result 代表其属性。在自定义的 resultMap 中第一列通常是主键 id，那么 id 和 result 有什么区别呢？id 和 result 都是映射单列值到一个属性或字段的简单数据类型。唯一不同的是，id 是作为唯一标识的，当与其他对象实例对比的时候，这个 id 很有用，尤其是应用到缓存和内嵌的结果映射。column 属性是指数据库的列名或者列标签别名，与数据库对应表中的列名称相同。property 属性是指映射数据库列的字段或属性，如果 JavaBean 的属性与给定的名称匹配，就会使用匹配的名字，否则 MyBatis 将搜索给定名称的字段。

实例见 7.4.1 节的实例 7.4 和 7.4.2 节的实例 7.5。

7.5.2 resultType

resultType 的属性值有三种：单一类型、组合类型、Map 类型。

第一种是单一类型，例如：

```
<select id="getCount" resultType="java.lang.Integer">
```

第二种是组合类型，一般都是 pojo 类，注意查询语句的列名或别名必须与 pojo 类的属性名称一致，否则无法映射，例如：

```
<select id="query2" resultType="com.icss.MyBatis.pojo.Student">
```

第三种是 Map 类型，列值会自动封装为键值对 Map 集合，键为列名，值为列值，例如：

```
<select id="query3" resultType="java.util.HashMap">
```

下面讲解这三种类型的实例。

【实例 7.6】 MyBatis 的结果类型：resultType。

使用 MyBatis 查询 Student 表中的数据。

准备工作：见实例 7.1 基于 MyBatis 的数据添加。

具体源代码如下所示（代码详见 MyBatis3_Ch07_Mybatis02\src\cn\mybatis\dao\ studentMapper.xml；MyBatis3_Ch07_Mybatis02\src\cn\mybatis\dao\StudentDao.java；MyBatis3_Ch07_Mybatis02\src\cn\mybatis\dao\impl\StudentDaoImpl.java；MyBatis3_Ch07_Mybatis02\src\cn\mybatis\domain\Student.java；MyBatis3_Ch07_Mybatis02\src\cn\mybatis\demo\Demo_01.java）。

（1）studentMapper.xml

```
<?xml version="1.0" encoding="UTF-8" ?>
<!DOCTYPE mapper PUBLIC "-//mybatis.org//DTD Mapper 3.0//EN"
"http://mybatis.org/dtd/mybatis-3-mapper.dtd">
<mapper namespace="Student">
   <resultMap type="cn.mybatis.domain.Student" id="studentResultMap">
       <id column="stu_id" property="stuId" jdbcType="INTEGER"
           javaType="java.lang.Integer" />
       <result column="stu_name" property="stuName" jdbcType="VARCHAR"
           javaType="java.lang.String" />
       <result column="stu_birthdate" property="stuBirthdate"
           jdbcType="DATE" javaType="java.util.Date" />
       <result column="stu_phone" property="stuPhone" jdbcType="VARCHAR"
           javaType="java.lang.String" />
   </resultMap>
   <!-- 返回总记录数 -->
   <select id="getCount" resultType="java.lang.Integer">
       select count(*) from student
   </select>
   <!-- 查询多条，返回类型为 pojo 类型，前提是列名（别名）必须和 pojo 类属性名称一致 -->
   <select id="query2" resultType="cn.mybatis.domain.Student">
        SELECT s.stu_id AS stuId,s.stu_name AS stuName,
        s.stu_birthdate AS stuBirthdate,s.stu_phone AS stuPhone
        FROM student s
   </select>
   <!-- 查询多条，返回类型为 List<Map>类型 -->
   <select id="query3" resultType="java.util.HashMap">
```

```
        select * from student
    </select>
</mapper>
```

（2）StudentDao.java

```
public int getCount();
public List<Student> selectStudent2();
public List selectStudent3();
```

（3）StudentDaoImpl.java

```
@Override
public int getCount() {
    SqlSession sess = null;
    try {
        sess = this.fac.openSession();
        SqlSession session = fac.openSession();
        Integer count = (Integer) session.selectOne("Student.getCount");
        sess.commit();
        return count;
    } catch (Exception e) {
        e.printStackTrace();
    } finally {
        sess.close();
    }
    return 0;
}
@Override
public List<Student> selectStudent2() {
    SqlSession sess = null;
    try {
        sess = fac.openSession();
        List<Student> list = sess.selectList("Student.query2");
        sess.commit();
        return list;
    } catch (Exception e) {
        e.printStackTrace();
    } finally {
        sess.close();
    }
    return null;
}
@Override
public List selectStudent3() {
    SqlSession sess = null;
    try {
        sess = fac.openSession();
        List list = sess.selectList("Student.query3");
        sess.commit();
        return list;
    } catch (Exception e) {
        e.printStackTrace();
    } finally {
        sess.close();
    }
    return null;
```

```
}
```

（4）Student.java

与实例 7.1 中的 Student.java 相同。

（5）Demo_01.java

```
@Test
@Test
public void testGetCount() {
    StudentDao dao=new StudentDaoImpl(fac);
    Integer count = dao.getCount();
    System.out.println("count=" + count);
}
@Test
public void testQuery2() {
    StudentDao dao=new StudentDaoImpl(fac);
    List<Student> list = dao.selectStudent2();
    for (Student stu : list) {
    System.out.println("testQuery2=="+stu);
    }
}
@Test
public void testQuery3() {
     StudentDao dao=new StudentDaoImpl(fac);
     List  list = dao.selectStudent3();
     for (int i=0;i<list.size();i++) {
     HashMap map=(HashMap)list.get(i);
     System.out.println("STU_ID=" +map.get("STU_ID"));
     System.out.println("STU_NAME=" + map.get("STU_NAME"));
     System.out.println("STU_BIRTHDATE=" + map.get("STU_BIRTHDATE"));
     System.out.println("STU_PHONE=" + map.get("STU_PHONE"));
     System.out.println("---------------------");
     }
}
```

运行程序。选中 Package Explorer 视图下 Demo_01.java 文件下的方法 testGetCount，单击鼠标右键，选择“Run As”菜单中的子菜单“2 Junit Test”，运行程序后控制台显示如图 7-7 所示。

```
count=2
```

图 7-7　查询 Student 表中的总记录数操作结果

运行程序。选中 Package Explorer 视图下 Demo_01.java 文件下的方法 testQuery2，单击鼠标右键，选择“Run As”菜单中的子菜单“2 Junit Test”，运行程序后控制台显示如图 7-8 所示。

```
testQuery2==Student [stuId=2, stuName=tom, stuBirthdate=Mon Dec 03 00:00:00 CST 2001, stuPhone=1505315****]
testQuery2==Student [stuId=3, stuName=tom, stuBirthdate=Mon Dec 03 00:00:00 CST 2001, stuPhone=1505315****]
```

图 7-8　查询 Student 表中的所有记录，返回值为 pojo 类型

运行程序。选中 Package Explorer 视图下 Demo_01.java 文件下的方法 testQuery3，单击

鼠标右键，选择“Run As”菜单中的子菜单“2 Junit Test”，运行程序后控制台显示如图 7-9 所示。

```
STU_ID=2
STU_NAME=tom
STU_BIRTHDATE=2001-12-03 00:00:00.0
STU_PHONE=1505315****
---------------------
STU_ID=3
STU_NAME=tom
STU_BIRTHDATE=2001-12-03 00:00:00.0
STU_PHONE=1505315****
---------------------
```

图 7-9　查询 Student 表中的所有记录，返回值为 List 类型

7.6　#和$的用法

#和$的用法

在 MyBatis 中，#和$的区别如下。

① #可以进行预编译，也可以进行类型匹配，#{变量名}会转化为 jdbc 的类型。

② $不能进行数据类型匹配，${变量名}可直接把${name}替换为 name 的内容。例如：

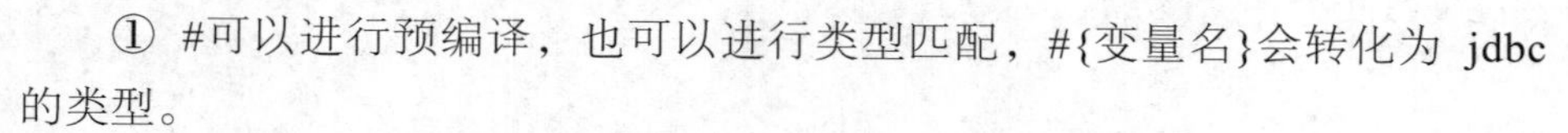

```
select * from tablename where id = #{id}
```

假设字段 id 的值为 12，如果数据库字段 id 为字符型，那么#{id}表示的就是'12'；如果字段 id 为整型，那么#{id}就是 12，会转化为 jdbc 的 select * from tablename where id=?，把参数?设置为 id 的值。

```
select * from tablename where id = ${id}
```

如果字段 id 为整型，SQL 语句就不会出错，但是如果字段 id 为字符型，那么 SQL 语句应该写成 select * from table where id = '${id}'。

③ 事实上在 MyBatis 中使用${id}这种标识符会直接抛异常，允许直接使用的标识符只能是${value}。

④ 如果不需要动态指定 SQL 语句，应该尽量使用#而不是$，因为$语法容易被 SQL 注入。

下面通过实例来讲解#和$的用法。

【实例 7.7】#和$的用法。

使用 MyBatis 查询 Student 表中的数据。

准备工作：见实例 7.1 基于 MyBatis 的数据添加。

具体源代码如下所示（代码详见 MyBatis3_Ch07_Mybatis02\src\cn\mybatis\dao\studentMapper.xml；MyBatis3_Ch07_Mybatis02\src\cn\mybatis\dao\StudentDao.java；MyBatis3_Ch07_Mybatis02\src\cn\mybatis\dao\impl\StudentDaoImpl.java；MyBatis3_Ch07_Mybatis02\src\cn\mybatis\domain\Student.java；MyBatis3_Ch07_Mybatis02\src\cn\mybatis\demo\Demo_01.java）。

（1）studentMapper.xml

```
<?xml version="1.0" encoding="UTF-8" ?>
<!DOCTYPE mapper PUBLIC "-//mybatis.org//DTD Mapper 3.0//EN"
"http://mybatis.org/dtd/mybatis-3-mapper.dtd">
```

```
<mapper namespace="Student">
   <resultMap type="cn.mybatis.domain.Student" id="studentResultMap">
       <id column="stu_id" property="stuId" jdbcType="INTEGER"
           javaType="java.lang.Integer" />
       <result column="stu_name" property="stuName" jdbcType="VARCHAR"
           javaType="java.lang.String" />
       <result column="stu_birthdate" property="stuBirthdate"
           jdbcType="DATE" javaType="java.util.Date" />
       <result column="stu_phone" property="stuPhone"
           jdbcType="VARCHAR" javaType="java.lang.String" />
   </resultMap>
   <!-- ${}的用法 -->
   <select id="query5" resultType="java.lang.Integer">
   select ${value} from student
   </select>
   <!-- #{}的用法 -->
   <select id="selectStudentByCondition" parameterType="cn.mybatis.domain.Student"
       resultMap="studentResultMap">
       select * from student where 1=1
       <!-- 注意 test 中 pojo 中的属性名称 -->
           <if test="stuName !=null and stuName != ''">
        <!-- 注意 like 的使用 -->
           and stu_name like '%'||#{stuName}||'%'
       </if>
       <if test="stuBirthdate !=null">
            <!-- 因 XML 文件中不能包含“<,>”这样的符号，因此是 cdata 区转换 -->
            <![CDATA[and stu_birthdate>#{stuBirthdate}]]>
       </if>
   </select>
</mapper>
```

（2）StudentDao.java

```
public Integer selectStudent5();
public List<Student> selectStudentByCondition(Student s);
```

（3）StudentDaoImpl.java

```
@Override
public Integer selectStudent5() {
    SqlSession sess = null;
    try {
        sess = this.fac.openSession();
        Integer s = sess.selectOne("Student.query5", "max(stu_id)");
        sess.commit();
        return s;
    } catch (Exception e) {
        e.printStackTrace();
    } finally {
        sess.close();
    }
    return null;
}
@Override
```

```
    public List<Student> selectStudentByCondition(Student s) {
        SqlSession sess = null;
        try {
            sess = this.fac.openSession();
            List<Student> list = sess.selectList("Student.selectStudentByCondition", s);
            sess.commit();
            return list;
        } catch (Exception e) {
            e.printStackTrace();
        } finally {
            sess.close();
        }
        return null;
    }
```

（4）Student.java

与实例 7.1 中的 Student.java 相同。

（5）Demo_01.java

```
    @Test
    public void testQuery5() {
        StudentDao dao=new StudentDaoImpl(fac);
        Integer result = dao.selectStudent5();
        System.out.println("result=" + result);
    }
    @Test
    public void selectByConditionTest() throws ParseException{
        Student s=new Student();
        s.setStuName("tom");
        s.setStuBirthdate(new SimpleDateFormat("yyyy-MM-dd").parse("2000-1-10"));
        StudentDao dao=new StudentDaoImpl(fac);
        List<Student> list=dao.selectStudentByCondition(s);
        System.out.println(list);
    }
```

运行程序。选择 Package Explorer 视图下 Demo_01.java 文件的方法 testQuery5，单击鼠标右键，选择“Run As”菜单中的子菜单“2 Junit Test”，运行程序后，控制台显示如图 7-10 所示。

```
result=3
```

图 7-10 查询 Student 表中的总记录数操作结果

运行程序。选择 Package Explorer 视图下 Demo_01.java 文件的方法 selectByConditionTest，单击鼠标右键，选择“Run As”菜单中的子菜单“2 Junit Test”，运行程序后，控制台显示如图 7-11 所示。

```
[Student [stuId=2, stuName=tom, stuBirthdate=Mon Dec 03 00:00:00 CST 2001, stuPhone=1505315****],
Student [stuId=3, stuName=tom, stuBirthdate=Mon Dec 03 00:00:00 CST 2001, stuPhone=1505315****]]
```

图 7-11 查询 Student 表中的所有记录，返回类型为 pojo 类型

本章小结

本章对基于 MyBatis 的增删改查操作、MyBatis 的结果类型、MyBatis 的参数类型、MyBatis 中#和$两种语法进行了较为详细的讲解，读者重点需要掌握的内容有 MyBatis 的增删改查操作、MyBatis 的结果类型和参数类型的用法等。本章的案例演示比较详细清晰，相信读者通过本章的学习就能对 MyBatis 进行初步的应用。

习题

1. 简述#{}和${}的区别。
2. MyBatis 的核心配置文件 sqlmapconfig.xml，除了常见的<select><insert><update><delete>标签外，还有哪些标签？

上机指导

表 tb_user 的表结构如图 7-12 所示。

名称	类型
ID	NUMBER(5)
NAME	VARCHAR2(20)
SEX	VARCHAR2(20)

图 7-12　表 tb_user 的表结构

1. 使用 MyBatis 框架模糊查询 tb_user 中 username 带“小”的用户信息（模糊查询）。
2. 使用 MyBatis 框架查询 tb_user 中 id=2 的用户。
3. 使用 MyBatis 框架添加 id=12 的用户到 tb_user 中。
4. 使用 MyBatis 框架修改 tb_user 中 id=12 的用户信息。
5. 使用 MyBatis 框架删除 tb_user 中 id=12 的用户信息。

08 第8章　MyBatis的动态SQL语句

学习目标

- 掌握动态 SQL 元素 if 的使用方法
- 掌握动态 SQL 元素 choose 的使用方法
- 掌握动态 SQL 元素 where 的使用方法
- 掌握动态 SQL 元素 trim 的使用方法
- 掌握动态 SQL 元素 foreach 的使用方法
- 掌握动态 SQL 元素 set 的使用方法

MyBatis 的强大特性之一便是它的动态 SQL。如果你有使用 JDBC 或其他类似框架的经验，你就能体会到根据不同条件拼接 SQL 语句的困难。例如，拼接时要确保添加必要的空格，还要注意去掉列表最后一个列名的逗号。使用动态 SQL 的特性可以彻底解决这种困难。MyBatis 提供了可以用在任意 SQL 映射语句中的强大的动态 SQL 语言以改变这种情况。MyBatis 的动态 SQL 是基于 OGNL 表达式的，它可以帮助我们方便地在 SQL 语句中实现某些逻辑。

8.1　if

if

在应用中，我们经常会使用一些动态的拼接条件。如果是 JDBC，我们可以用程序来拼接 SQL 语句；如果是 MyBatis，我们就可以使用动态 SQL 语句。例如，按照员工的姓名和工资来搜索员工信息，如果姓名和工资的检索值为空，则忽略这个检索条件。一般来说，我们都会用 where 1=1 这类写法来实现，但是 MyBatis 就需要动态语句来实现。下面先介绍动态 SQL 元素 if 的使用。

【实例 8.1】动态 SQL 元素 if 的使用案例。

使用动态 SQL 元素 if 查询 Student 表中符合条件的记录。

准备工作：见实例 7.1 基于 MyBatis 的数据添加。

具体源代码如下所示（代码详见MyBatis3_Ch08_Mybatis03\src\cn\mybatis\dao\studentMapper.xml；MyBatis3_Ch08_Mybatis03\src\cn\mybatis\dao\StudentDao.java；MyBatis3_Ch08_Mybatis03\src\cn\mybatis\dao\impl\StudentDaoImpl.java；MyBatis3_Ch08_Mybatis03\src\cn\mybatis\domain\Student.java；MyBatis3_Ch08_Mybatis03\src\cn\mybatis\demo\Demo_01.java）。

（1）studentMapper.xml

```
<?xml version="1.0" encoding="UTF-8" ?>
<!DOCTYPE mapper PUBLIC "-//mybatis.org//DTD Mapper 3.0//EN"
"http://mybatis.org/dtd/mybatis-3-mapper.dtd">
<mapper namespace="Student">
<resultMap type="cn.mybatis.domain.Student" id="studentResultMap">
        <id column="stu_id" property="stuId" jdbcType="INTEGER"
            javaType="java.lang.Integer" />
        <result column="stu_name" property="stuName" jdbcType="VARCHAR"
            javaType="java.lang.String" />
        <result column="stu_birthdate" property="stuBirthdate"
            jdbcType="DATE" javaType="java.util.Date" />
        <result column="stu_phone" property="stuPhone" jdbcType= "VARCHAR"
            javaType="java.lang.String" />
</resultMap>
<select id="queryByCondition1" parameterType="cn.mybatis.domain.Student"
   resultMap="studentResultMap">
select * from student where 1=1
<if test="stuName != null and stuName != ''">
   and stu_name=#{stuName}
</if>
</select>
</mapper>
```

（2）StudentDao.java

```
public List<Student> queryByCondition1(Student s);
```

（3）StudentDaoImpl.java

```
@Override
public List<Student> queryByCondition1(Student s) {
    SqlSession sess = null;
    try {
        sess = this.fac.openSession();
        List<Student> list = sess.selectList("Student.queryByCondition1", s);
        sess.commit();
        return list;
    } catch (Exception e) {
        e.printStackTrace();
    } finally {
        sess.close();
    }
    return null;
}
```

（4）Student.java

与实例7.1中的Student.java相同。

（5）Demo_01.java

```
@Test
public void queryByCondition1() throws ParseException{
    Student s=new Student();
    s.setStuName("tom");
    StudentDao dao=new StudentDaoImpl(fac);
    List<Student> list=dao.queryByCondition2(s);
    System.out.println(list);
}
```

运行程序。选中 Package Explorer 视图下 Demo_01.java 文件的方法 queryByCondition1，单击鼠标右键，选择“Run As”菜单中的子菜单“2 Junit Test”，运行程序后查询结果如图 8-1 所示。

```
[Student [stuId=22, stuName=tom, stuBirthdate=Mon Dec 03 00:00:00 CST 2001, stuPhone=1505315****],
Student [stuId=2, stuName=tom, stuBirthdate=Sat Nov 03 00:00:00 CST 1990, stuPhone=1505315****]]
```

图 8-1　动态 SQL 元素 if 的使用案例

【案例解析】

注意 if 元素中的查询条件：当传入的属性值为空或空字符串时，将会忽略掉条件。

8.2　choose

choose

有时我们不想应用所有的条件语句，只想从中选择其一。MyBatis 中的 choose 元素的作用相当于 Java 中的 if…else if…else 语句或 switch 语句。choose 元素按顺序判断其内部<when>标签中的 test 条件是否成立，如果有一个成立，则 choose 结束。当 choose 中所有<when>的条件都不满足时，则执行 otherwise 中的 SQL。类似于 Java 的 switch 语句，choose 为 switch，when 为 case，otherwise 则为 default。

【实例 8.2】动态 SQL 元素 choose 的使用案例。

使用 MyBatis 中的动态 SQL 元素 choose 查询 Student 表中符合姓名、生日、电话查询条件的记录。

准备工作：见实例 7.1 基于 MyBatis 的数据添加。

具体源代码如下所示（代码详见 MyBatis3_Ch08_Mybatis03\src\cn\mybatis\dao\studentMapper.xml；MyBatis3_Ch08_Mybatis03\src\cn\mybatis\dao\StudentDao.java；MyBatis3_Ch08_Mybatis03\src\cn\mybatis\dao\impl\StudentDaoImpl.java；MyBatis3_Ch08_Mybatis03\src\cn\mybatis\domain\Student.java；MyBatis3_Ch08_Mybatis03\src\cn\mybatis\demo\Demo_01.java）。

（1）studentMapper.xml

```
<?xml version="1.0" encoding="UTF-8" ?>
<!DOCTYPE mapper PUBLIC "-//mybatis.org//DTD Mapper 3.0//EN"
"http://mybatis.org/dtd/mybatis-3-mapper.dtd">
<mapper namespace="Student">
<resultMap type="cn.mybatis.domain.Student" id="studentResultMap">
```

```
    <id column="stu_id" property="stuId" jdbcType="INTEGER"
        javaType="java.lang.Integer" />
    <result column="stu_name" property="stuName" jdbcType="VARCHAR"
        javaType="java.lang.String" />
    <result column="stu_birthdate" property="stuBirthdate"
        jdbcType="DATE" javaType="java.util.Date" />
    <result column="stu_phone" property="stuPhone"
        jdbcType="VARCHAR" javaType="java.lang.String" />
</resultMap>
<select id="queryByCondition2" parameterType="cn.mybatis.domain.Student"
    resultMap="studentResultMap">
select * from student where 1=1
<choose>
<when test="stuName != null and stuName != ''">
    and stu_name=#{stuName}
</when>
<when test="stuBirthdate != null">
    and stu_birthdate=#{stuBirthdate}
</when>
<otherwise>
    and stu_phone=#{stuPhone}
</otherwise>
</choose>
</select>
</mapper>
```

（2）StudentDao.java

```
public List<Student> queryByCondition2(Student s);
```

（3）StudentDaoImpl.java

```
@Override
public List<Student> queryByCondition2(Student s) {
    SqlSession sess = null;
    try {
        sess = this.fac.openSession();
        List<Student> list = sess.selectList("Student.queryByCondition2", s);
        sess.commit();
        return list;
    } catch (Exception e) {
        e.printStackTrace();
    } finally {
        sess.close();
    }
    return null;
}
```

（4）Student.java

与实例 7.1 中的 Student.java 相同。

（5）Demo_01.java

```
@Test
public void queryByCondition2() throws ParseException{
    Student s=new Student();
    s.setStuName("tom");
```

```
        s.setStuBirthdate(new SimpleDateFormat("yyyy-MM-dd").parse("2000-1-10"));
        s.setStuPhone("18653186541");
        StudentDao dao=new StudentDaoImpl(fac);
        List<Student> list=dao.queryByCondition2(s);
        System.out.println(list);
    }
```

运行程序。选中 Package Explorer 视图下 Demo_01.java 文件下的方法 queryByCondition2，单击鼠标右键，选择“Run As”菜单中的子菜单“2 Junit Test”，运行程序后查询结果如图 8-2 所示。

```
[Student [stuId=22, stuName=tom, stuBirthdate=Mon Dec 03 00:00:00 CST 2001, stuPhone=1505315****],
Student [stuId=2, stuName=tom, stuBirthdate=Sat Nov 03 00:00:00 CST 1990, stuPhone=1505315****]]
```

图 8-2　动态 SQL 元素 choose 的使用案例

8.3　where

MyBatis 中动态 SQL 元素 where 也可以设置 SQL 的查询条件，它会自动根据条件的个数增删 where 语句和 and 运算符，所以不需要写 where 1=1 之类的语句。

【实例 8.3】动态 SQL 元素 where 的使用案例。

使用 MyBatis 中的动态 SQL 元素 where 查询 Student 表中符合姓名、生日、电话查询条件的记录。

准备工作：见实例 7.1 基于 MyBatis 的数据添加。

具体源代码如下所示（代码详见 MyBatis3_Ch08_Mybatis03\src\cn\mybatis\dao\studentMapper.xml；MyBatis3_Ch08_Mybatis03\src\cn\mybatis\dao\StudentDao.java；MyBatis3_Ch08_Mybatis03\src\cn\mybatis\dao\impl\StudentDaoImpl.java；MyBatis3_Ch08_Mybatis03\src\cn\mybatis\domain\Student.java；MyBatis3_Ch08_Mybatis03\src\cn\mybatis\demo\Demo_01.java）。

（1）studentMapper.xml

```
<?xml version="1.0" encoding="UTF-8" ?>
<!DOCTYPE mapper PUBLIC "-//mybatis.org//DTD Mapper 3.0//EN"
"http://mybatis.org/dtd/mybatis-3-mapper.dtd">
<mapper namespace="Student">
<resultMap type="cn.mybatis.domain.Student" id="studentResultMap">
    <id column="stu_id" property="stuId" jdbcType="INTEGER"
        javaType="java.lang.Integer" />
    <result column="stu_name" property="stuName" jdbcType="VARCHAR"
        javaType="java.lang.String" />
    <result column="stu_birthdate" property="stuBirthdate"
        jdbcType="DATE" javaType="java.util.Date" />
    <result column="stu_phone" property="stuPhone" jdbcType="VARCHAR"
        javaType="java.lang.String" />
</resultMap>
<select id="queryByCondition3" parameterType="cn.mybatis.domain.Student"
    resultMap="studentResultMap">
```

```
    select * from student
    <where>
    <if test="stuName != null and stuName != ''">
        and stu_name=#{stuName}
    </if>
    <if test="stuBirthdate != null">
        and stu_birthdate=#{stuBirthdate}
    </if>
    <if test="stuPhone != null and stuPhone != ''">
        and stu_phone=#{stuPhone}
    </if>
    </where>
    </select>
    </mapper>
```

（2）StudentDao.java

```
    public List<Student> queryByCondition3(Student s);
```

（3）StudentDaoImpl.java

```
    @Override
    public List<Student> queryByCondition3(Student s) {
        SqlSession sess = null;
        try {
            sess = this.fac.openSession();
            List<Student> list = sess.selectList("Student.queryByCondition3", s);
            sess.commit();
            return list;
        } catch (Exception e) {
            e.printStackTrace();
        } finally {
            sess.close();
        }
        return null;
    }
```

（4）Student.java

与实例7.1中的Student.java相同。

（5）Demo_01.java

```
    @Test
    public void queryByCondition3() throws ParseException{
        Student s=new Student();
        s.setStuName("tom");
        s.setStuBirthdate(new SimpleDateFormat("yyyy-MM-dd").parse("2001-12-3"));
        s.setStuPhone("1505315****");
        StudentDao dao=new StudentDaoImpl(fac);
        List<Student> list=dao.queryByCondition3(s);
        System.out.println(list);
    }
```

运行程序。选中Package Explorer视图下Demo_01.java文件下的方法queryByCondition3，单击鼠标右键，选择“Run As”菜单中的子菜单“2 Junit Test”，运行程序后查询结果如图8-3所示。

```
[Student [stuId=22, stuName=tom, stuBirthdate=Mon Dec 03 00:00:00 CST 2001, stuPhone=1505315****]]
```

图 8-3 动态 SQL 元素 where 的使用案例

8.4 trim

trim

MyBatis 中的动态 SQL 元素 trim 的主要功能是可以在自己包含的内容前加上某些前缀，也可以在其后加上某些后缀，与之对应的属性是 prefix 和 suffix；可以把首部的某些内容覆盖（即忽略），也可以把尾部的某些内容覆盖，与之对应的属性是 prefixOverrides 和 suffixOverrides。

【实例 8.4】动态 SQL 元素 trim 的使用案例。

使用 MyBatis 中的动态 SQL 元素 trim 查询 Student 表中符合姓名、生日、电话查询条件的记录。

准备工作：见实例 7.1 基于 MyBatis 的数据添加。

具体源代码如下所示（代码详见 MyBatis3_ch08_Mybatis03\src\cn\mybatis\dao\studentMapper.xml；\cn\mybatis\dao\StudentDao.java；MyBatis3_ch08_Mybatis03\src\cn\mybatis\dao\impl\StudentDaoImpl.java；MyBatis3_ch08_Mybatis03\src\cn\mybatis\domain\Student.java；MyBatis3_ch08_Mybatis03\src\cn\mybatis\demo\Demo_01.java）。

（1）studentMapper.xml

```
<?xml version="1.0" encoding="UTF-8" ?>
<!DOCTYPE mapper PUBLIC "-//mybatis.org//DTD Mapper 3.0//EN"
"http://mybatis.org/dtd/mybatis-3-mapper.dtd">
<mapper namespace="Student">
<resultMap type="cn.mybatis.domain.Student" id="studentResultMap">
    <id column="stu_id" property="stuId" jdbcType="INTEGER"
        javaType="java.lang.Integer" />
    <result column="stu_name" property="stuName" jdbcType="VARCHAR"
        javaType="java.lang.String" />
    <result column="stu_birthdate" property="stuBirthdate"
        jdbcType="DATE" javaType="java.util.Date" />
    <result column="stu_phone" property="stuPhone"
        jdbcType="VARCHAR" javaType="java.lang.String" />
</resultMap>
<select id="queryByCondition4" parameterType="cn.mybatis.domain.Student"
    resultMap="studentResultMap">
select * from student
<trim prefix="where" prefixOverrides="and|or">
<if test="stuName != null and stuName != ''">
    and stu_name like '%'||#{stuName}||'%'
</if>
<if test="stuBirthdate != null">
    and stu_birthdate=#{stuBirthdate}
</if>
<if test="stuPhone != null and stuPhone != ''">
    or stu_phone=#{stuPhone}
</if>
</trim>
```

```
</select>
</mapper>
```

（2）StudentDao.java

```
public List<Student> queryByCondition4(Student s);
```

（3）StudentDaoImpl.java

```
@Override
public List<Student> queryByCondition4(Student s) {
    SqlSession sess = null;
    try {
        sess = this.fac.openSession();
        List<Student> list = sess.selectList("Student.queryByCondition4", s);
        sess.commit();
        return list;
    } catch (Exception e) {
        e.printStackTrace();
    } finally {
        sess.close();
    }
    return null;
}
```

（4）Student.java

与实例 7.1 中的 Student.java 相同。

（5）Demo_01.java

```
@Test
public void queryByCondition4() throws ParseException{
    Student s=new Student();
    s.setStuName("tom");
    s.setStuBirthdate(new SimpleDateFormat("yyyy-MM-dd").parse("2001-12-3"));
    s.setStuPhone("1505315****");
    StudentDao dao=new StudentDaoImpl(fac);
    List<Student> list=dao.queryByCondition4(s);
    System.out.println(list);
}
```

运行程序。选中 Package Explorer 视图下 Demo_01.java 文件下的方法 queryByCondition4，单击鼠标右键，选择“Run As”菜单中的子菜单“2 Junit Test”，运行程序后查询结果如图 8-4 所示。

```
[Student [stuId=22, stuName=tom, stuBirthdate=Mon Dec 03 00:00:00 CST 2001, stuPhone=1505315****],
Student [stuId=2, stuName=tom, stuBirthdate=Sat Nov 03 00:00:00 CST 1990, stuPhone=1505315****]]
```

图 8-4　动态 SQL 元素 trim 的使用案例

8.5　foreach

foreach

foreach 主要用在构建 in 条件中，它可以在 SQL 语句中迭代一个集合。foreach 元素的属性主要有 item、index、collection、open、separator、close。item

表示集合中每一个元素进行迭代时的别名；index 指定一个名字，用于表示在迭代过程中，每次迭代到的位置；open 表示该语句以什么开始；separator 表示在每次进行迭代之间以什么符号作为分隔符；close 表示以什么结束；在使用 foreach 的时候，最关键也最容易出错的就是 collection 属性，该属性是必须指定的，但在不同情况下该属性的值是不一样的，主要有以下 3 种情况。

（1）如果传入的是单参数且参数类型是一个 List 的时候，collection 属性值为 list。

（2）如果传入的是单参数且参数类型是一个 array 数组的时候，collection 的属性值为 array。

（3）如果传入的参数是多个时，则需要把它们封装成一个 Map，当然单参数也可以封装成 Map，实际上如果传入参数，在 MyBatis 里面也会把它封装成一个 Map 的，Map 的 key 就是参数名，所以这个时候，collection 属性值就是传入的 List 或 array 对象在自己封装的 Map 里面的 key。

【实例 8.5】动态 SQL 元素 foreach 的使用案例。

使用 MyBatis 中的动态 SQL 元素 foreach 查询 Student 表中符合某些查询条件的记录。

准备工作：见实例 7.1 基于 MyBatis 的数据添加。

具体源代码如下所示（代码详见 MyBatis3_Ch08_Mybatis03\src\cn\mybatis\dao\studentMapper.xml；MyBatis3_Ch08_Mybatis03\src\cn\mybatis\dao\StudentDao.java；MyBatis3_Ch08_Mybatis03\src\cn\mybatis\dao\impl\StudentDaoImpl.java；MyBatis3_Ch08_Mybatis03\src\cn\mybatis\domain\Student.java；MyBatis3_Ch08_Mybatis03\src\cn\mybatis\demo\Demo_01.java）。

（1）studentMapper.xml

```
<?xml version="1.0" encoding="UTF-8" ?>
<!DOCTYPE mapper PUBLIC "-//mybatis.org//DTD Mapper 3.0//EN"
"http://mybatis.org/dtd/mybatis-3-mapper.dtd">
<mapper namespace="Student">
<resultMap type="cn.mybatis.domain.Student" id="studentResultMap">
    <id column="stu_id" property="stuId" jdbcType="INTEGER"
        javaType="java.lang.Integer" />
    <result column="stu_name" property="stuName" jdbcType="VARCHAR"
        javaType="java.lang.String" />
    <result column="stu_birthdate" property="stuBirthdate"
        jdbcType="DATE" javaType="java.util.Date" />
    <result column="stu_phone" property="stuPhone" jdbcType="VARCHAR"
        javaType="java.lang.String" />
</resultMap>
<!-- 动态 SQL：传入 array 数组 -->
<select id="queryByInArray" resultMap="studentResultMap">
   select * from student
<if test="array.length>0">
where stu_id in
<foreach collection="array" index="i" item="stuId" open="(" close=")"
separator=",">
#{stuId}
</foreach>
</if>
</select>
<!-- 动态 SQL：传入 List 集合 -->
<select id="queryByInList" resultMap="studentResultMap">
select * from student
<if test="list.size()>0">
```

```
    where stu_id in
    <foreach collection="list" index="i" item="stuId" open="(" close=")"
        separator=",">
        #{stuId}
    </foreach>
    </if>
    </select>
    <!-- 动态 SQL：传入 Map 集合包含 List 集合 -->
    <select id="queryByInMap" resultMap="studentResultMap">
    select * from student
    <if test="ids.size()>0">
    where stu_id in
    <foreach collection="ids" index="i" item="stuId" open="(" close=")"
        separator=",">
    #{stuId}
    </foreach>
    </if>
    </select>
    </mapper>
```

（2）StudentDao.java

```
    public List<Student> queryByInArray(int[] ids);
    public List<Student> queryByInList(List ids);
    public List<Student> queryByInMap(Map map);
```

（3）StudentDaoImpl.java

```
    @Override
    public List<Student> queryByInArray(int[] ids) {
        SqlSession sess = null;
        try {
            sess = this.fac.openSession();
            List<Student> list = sess.selectList("Student.queryByInArray", ids);
            sess.commit();
            return list;
        } catch (Exception e) {
            e.printStackTrace();
        } finally {
            sess.close();
        }
        return null;
    }
    @Override
    public List<Student> queryByInList(List ids) {
        SqlSession sess = null;
        try {
            sess = this.fac.openSession();
            List<Student> list = sess.selectList("Student.queryByInList", ids);
            sess.commit();
            return list;
        } catch (Exception e) {
            e.printStackTrace();
        } finally {
            sess.close();
        }
```

```
            return null;
    }
    @Override
    public List<Student> queryByInMap(Map map) {
        SqlSession sess = null;
        try {
            sess = this.fac.openSession();
            List<Student> list = sess.selectList("Student.queryByInMap", map);
            sess.commit();
            return list;
        } catch (Exception e) {
            e.printStackTrace();
        } finally {
            sess.close();
        }
        return null;
    }
```

（4）Student.java

与实例 7.1 中的 Student.java 相同。

（5）Demo_01.java

```
    @Test
    public void queryByInArray() throws ParseException{
        StudentDao dao=new StudentDaoImpl(fac);
        int[] ids = new int[]{2,3,4};
        List<Student> list=dao.queryByInArray(ids);
        for (Student item : list) {
            System.out.println(item);
            }
    }
    @Test
    public void queryByInList() throws ParseException{
        StudentDao dao=new StudentDaoImpl(fac);
        List ids = new ArrayList();
        ids.add(2);
        ids.add(3);
        ids.add(4);
        List<Student> list=dao.queryByInList(ids);
        for (Student item : list) {
            System.out.println(item);
            }
    }
    @Test
    public void queryByInMap() throws ParseException{
        StudentDao dao=new StudentDaoImpl(fac);
        List ids = new ArrayList();
        ids.add(2);
        ids.add(3);
        ids.add(4);
        Map map = new HashMap();
        map.put("ids", ids);
        List<Student> list=dao.queryByInMap(map);
        for (Student item : list) {
            System.out.println(item);
```

```
        }
    }
```

运行程序。选中 Package Explorer 视图下 Demo_01.java 文件中的方法 queryByInArray，单击鼠标右键，选择“Run As”菜单中的子菜单“2 Junit Test”，运行程序后查询结果如图 8-5 所示。

```
Student [stuId=2, stuName=tom, stuBirthdate=Sat Nov 03 00:00:00 CST 1990, stuPhone=1505315****]
```

图 8-5　动态 SQL 元素 foreach 的使用案例

执行 queryByInArray 方法可测试 foreach 元素中传入的是单参数且参数类型是一个 List。而执行 queryByInList 方法可测试 foreach 元素中传入的是单参数且参数类型是一个 array 数组，其执行结果和图 8-5 相同。执行 queryByInMap()方法可测试 foreach 元素中传入的是单参数且参数类型是一个 Map，执行结果也和图 8-5 相同。

8.6　set

set

set 元素主要用在更新操作的时候，它的主要功能和 where 元素其实是差不多的，主要是在包含的语句前输出一个 set，如果包含的语句是以逗号结束的话将会把该逗号忽略，如果 set 包含的内容为空的话则会出错。有了 set 元素就可以动态更新那些修改了的字段。

【实例 8.6】动态 SQL 元素 set 的使用案例。

使用 MyBatis 中的动态 SQL 元素 set 更新 Student 表中的记录。

准备工作：见实例 7.1 基于 MyBatis 的数据添加。

具体源代码如下所示（代码详见 MyBatis3_Ch08_Mybatis03\src\cn\mybatis\dao\studentMapper.xml；MyBatis3_Ch08_Mybatis03\src\cn\mybatis\dao\StudentDao.java；MyBatis3_Ch08_Mybatis03\src\cn\mybatis\dao\impl\StudentDaoImpl.java；MyBatis3_Ch08_Mybatis03\src\cn\mybatis\domain\Student.java；MyBatis3_Ch08_Mybatis03\src\cn\mybatis\demo\Demo_01.java）。

（1）studentMapper.xml

```
<?xml version="1.0" encoding="UTF-8" ?>
<!DOCTYPE mapper PUBLIC "-//mybatis.org//DTD Mapper 3.0//EN"
"http://mybatis.org/dtd/mybatis-3-mapper.dtd">
<mapper namespace="Student">
<!-- 动态 SQL: set 更新 -->
<update id="updateByCondition" parameterType="cn.mybatis.domain.Student">
update student
<set>
<if test="stuName!=null and stuName!=''">
stu_name=#{stuName},
</if>
<if test="stuBirthdate!=null">
stu_birthdate=#{stuBirthdate},
</if>
<if test="stuPhone!=null and stuPhone!=''">
stu_phone=#{stuPhone}
```

```
</if>
</set>
where stu_id=#{stuId}
</update>
</mapper>
```

（2）StudentDao.java

```
public void updateByCondition(Student s);
```

（3）StudentDaoImpl.java

```
@Override
public  void updateByCondition(Student s) {
     SqlSession sess = null;
     try {
          sess = this.fac.openSession();
          sess.update("Student.updateByCondition", s);
          sess.commit();
     } catch (Exception e) {
          e.printStackTrace();
     } finally {
          sess.close();
     }
}
```

（4）Student.java

与实例 7.1 中的 Student.java 相同。

（5）Demo_01.java

```
@Test
public void updateByCondition() throws Exception{
     Student s=new Student();
     s.setStuId(1);
     s.setStuName("Alice");
     s.setStuBirthdate(new SimpleDateFormat("yyyy-MM-dd").parse("2000-12-3"));
     s.setStuPhone("1385318****");
     StudentDao dao=new StudentDaoImpl(fac);
     dao.updateByCondition(s);
}
```

运行程序。选中 Package Explorer 视图下 Demo_01.java 文件中的方法 updateByCondition，单击鼠标右键，选择“Run As”菜单中的子菜单“2 Junit Test”，运行程序后，Student 表中修改后的该条记录显示如图 8-6 所示。

STU_ID	STU_NAME	STU_BIRTHDATE	STU_PHONE
1	Alice ...	2000/12/3 ▾	1385318**** ...

图 8-6　动态 SQL 元素 set 的使用案例

8.7　<SQL>和<include>

<SQL>和<include>

在 MyBatis 中可以编写一些语句片段<SQL>标签，然后用<include>引用。<SQL>用来封装 SQL 语句，<include>来调用，这样可以使 SQL 语句片段得

到重用。

【实例 8.7】动态 SQL 元素中的<SQL>和<include>的使用案例。

使用 MyBatis 中的动态 SQL 元素<SQL>和<include>查询 Student 表中的记录和记录数。

准备工作：见实例 7.1 基于 MyBatis 的数据添加。

具体源代码如下所示（代码详见 MyBatis3_Ch08_Mybatis03\src\cn\mybatis\dao\studentMapper.xml；MyBatis3_Ch08_Mybatis03\src\cn\mybatis\dao\StudentDao.java；MyBatis3_Ch08_Mybatis03\src\cn\mybatis\dao\impl\StudentDaoImpl.java；MyBatis3_Ch08_Mybatis03\src\cn\mybatis\domain\Student.java；MyBatis3_Ch08_Mybatis03\src\cn\mybatis\demo\Demo_01.java）。

（1）studentMapper.xml

```
<?xml version="1.0" encoding="UTF-8" ?>
<!DOCTYPE mapper PUBLIC "-//mybatis.org//DTD Mapper 3.0//EN"
    "http://mybatis.org/dtd/mybatis-3-mapper.dtd">
<mapper namespace="Student">
<sql id="SQL_select">
select *
</sql>
<sql id="SQL_count">
select count(*)
</sql>
    <!-- 包含 SQL 片段 -->
    <select id="query6" resultMap="studentResultMap">
        <include refid="SQL_select"/>
        from student
    </select>
    <select id="query7" resultType="java.lang.Integer">
        <include refid="SQL_count"/>
        from student
    </select>
</mapper>
```

（2）StudentDao.java

```
public List<Student> query6();
public int query7();
```

（3）StudentDaoImpl.java

```
@Override
public List<Student> query6() {
    SqlSession sess = null;
    try {
        sess = this.fac.openSession();
        List<Student> list = sess.selectList("Student.query6");
        sess.commit();
        return list;
    } catch (Exception e) {
        e.printStackTrace();
    } finally {
        sess.close();
    }
    return null;
```

```
    }
    @Override
    public int query7() {
        SqlSession sess = null;
        try {
            sess = this.fac.openSession();
            Integer num= sess.selectOne("Student.query7");
            sess.commit();
            return num;
        } catch (Exception e) {
            e.printStackTrace();
        } finally {
            sess.close();
        }
        return 0;
    }
```

（4）Student.java

与实例 7.1 中的 Student.java 相同。

（5）Demo_01.java

```
    @Test
    public void query6(){
        StudentDao dao=new StudentDaoImpl(fac);
        List<Student> list=dao.query6();
        System.out.println(list);
    }
    @Test
    public void query7(){
        StudentDao dao=new StudentDaoImpl(fac);
        int count=dao.query7();
        System.out.println(count);
    }
```

运行程序。选中 Package Explorer 视图下 Demo_01.java 文件下的方法 query6，单击鼠标右键，选择“Run As”菜单中的子菜单“2 Junit Test”，运行程序后查询出了 Student 表中的数据如图 8-7 所示。

```
[Student [stuId=1, stuName=Alice, stuBirthdate=Sun Dec 03 00:00:00 CST 2000, stuPhone=1385318**** ],
Student [stuId=22, stuName=tom, stuBirthdate=Mon Dec 03 00:00:00 CST 2001, stuPhone=1505315**** ],
Student [stuId=2, stuName=tom, stuBirthdate=Sat Nov 03 00:00:00 CST 1990, stuPhone=1505315**** ]]
```

图 8-7　动态 SQL 元素<SQL>和<include>的使用案例

选中 Package Explorer 视图下 Demo_01.java 文件下的方法 query7，单击鼠标右键，选择“Run As”菜单中的子菜单“2 Junit Test”，运行程序后，查询出 Student 表中的记录个数，如图 8-8 所示。

```
3
```

图 8-8　动态 SQL 元素<SQL>和<include>的使用案例

【案例解析】

这个案例查询了 Student 表中的记录和记录数。它们的代码的不同在于 studentMapper.xml 中的 SQL 元素内容不同，StudentDaoImpl.java 中获取查询结果使用的 SqlSession 类的方法不同。

本章小结

本章讲解了 MyBatis 中的动态 SQL 元素（if、choose、trim、where、set、foreach）的使用方法，还讲解了在动态 SQL 中如何使用<SQL>标签和<include>标签抽取可重用的 SQL 片段。相信读者通过学习 MyBatis 中动态 SQL 的操作可以极大地简化拼装 SQL 的操作。

习题

1. 简述 MyBatis 框架动态 SQL 中的主要元素及其功能。
2. 简述 MyBatis 框架动态 SQL 中的 foreach 元素的 collection 属性的注意事项。

上机指导

1. 使用 MyBatis 框架查询 user 表中的总记录数（要求使用 if、where）。
2. 使用 MyBatis 框架查询 user 表中 id=8 及 id=10 的用户信息（要求使用 foreach）。

09 第9章 使用MyBatis动态代理技术实现DAO接口

学习目标

- 了解 MyBatis 动态代理技术的基本原理
- 掌握使用 MyBatis 动态代理技术实现 DAO 接口的 CRUD 操作方法

MyBatis 框架为用户提供了稳定且方便的数据库访问功能，极大地规范了数据访问的代码开发。为了更加方便企业级项目的开发，节约开发时间和开发成本，MyBatis 还支持使用动态代理直接实现 DAO 接口，本章主要讨论 MyBatis 动态代理技术的实现。

9.1 MyBatis 动态代理的概念

MyBatis 动态代理的概念

在介绍 MyBatis 的动态代理之前，先来看使用 MyBatis 进行传统的 DAO 访问，对数据库中的学生信息进行添加的操作，具体代码如下。

```
public class StudentDao implements StudentMapper{
    public void insertStu(Stu s) {
        InputStream is=null;
        try {
            is = Resources.getResourceAsStream("mybatisConfig.
        xml");
        } catch (IOException e) {
            e.printStackTrace();
        }
    SqlSessionFactory ssf=new SqlSessionFactoryBuilder().
        build(is);
        SqlSession ss=ssf.openSession();
        ss.insert("stus.insertStu", s);
        ss.commit();
        ss.close();
    }
}
```

在测试类中进行测试，代码如下。

```
public class Demo {
```

```
    /**
     * 通过普通 DAO 实现类访问数据库
     */
    publicstaticvoid main(String[] args) {
        Stu s=new Stu();
        s.setId(102);
        s.setName("mary");
        s.setAge(18);
        new StudentDao().insertStu(s);
    }
}
```

在使用 DAO 类实现接口后，我们会通过 SqlSession 对象执行增删改查操作（即人们常说的 CRUD，Create 增加、Retrieve 检索、Update 更改、Delete 删除）。在该类以及其他 DAO 实现类中，可能还有大量其他的 CRUD 操作，而这些操作的执行方式都是相同的。这样就给用户的代码开发带来了大量的重复工作，降低了开发效率。

为提高开发效率，针对这些重复工作，MyBatis 框架为用户提供了一种简便的数据访问方式，即直接通过配置文件，按照一定的规则进行与接口的映射，自动实现对数据库的访问，从而无须进行 DAO 实现类的开发。接口声明完成以后只需要在使用的时候通过 SqlSession 对象的 getMapper 方法，自动获取对应 DAO 接口的代理对象来完成数据访问。在这个过程中，框架通过 MapperMethod 类（整个代理机制的核心类），对 SqlSession 中的操作进行了封装使用，与我们直接用 SqlSession 对象调用相应的 CRUD 的方法是一样的。

9.2　动态代理实现插入操作

动态代理实现插入操作

MyBatis 框架要完成动态代理实现 DAO 接口，必须按照一定的规则进行配置。

MyBatis 动态代理生成 DAO 的步骤如下。

（1）编写数据管理的接口 XxxMapper.java。

（2）编写接口对应的配置文件 XxxMapper.xml。

① namespace 必须和 DAO 接口的全路径保持一致（即物理路径的文件夹名称相同）。

② statement 的 id 必须和 DAO 接口的方法名保持一致。

③ statement 的 resultType 类型必须和方法返回值类型保持一致。

（3）通过 sqlSession.getMapper（类的字节码对象）获取代理之后的 Mapper 实现类对象。

【实例 9.1】动态代理实现 DAO 接口的使用案例。

（1）设置 mapper 文件中的 namespace 属性值与接口名一致。

```
<mappernamespace="com.inspur.service.StudentMapper">

package com.inspur.service;
import com.inspur.pojo.Stu;
public interface StudentMapper {
    voidinsertStu(Stu s);
```

```
    }
```

（2）设置 mapper 文件中<insert>标签的 id 属性值与接口中对应的方法名一致。

```
<mapper namespace="com.inspur.service.StudentMapper">
    <insert id="insertStu" parameterType="com.inspur.pojo.Stu">
        insert into stus_mybatis_ch09 values(#{id},#{name},#{age})
    </insert>
</mapper>

public interface StudentMapper {
    void insertStu(Student s);
}
```

（3）在测试类中通过 SqlSession 接口的 getMapper 方法获取 mapper 对象，并执行 CRUD 操作，省略了 DAO 实现类的开发。

```
public class Demo2 {
    /**
     * 进行动态代理访问数据库
     */
    public staticvoid main(String[] args) {
        InputStream is=null;
        try {
            is = Resources.getResourceAsStream("mybatisConfig.xml");
        } catch (IOException e) {
            e.printStackTrace();
        }
        SqlSessionFactory ssf=new SqlSessionFactoryBuilder().build(is);
        SqlSession ss=ssf.openSession();
        //根据接口参数获取 mapper 对象，直接进行 CRUD 操作
        StudentMapper mapper=ss.getMapper(StudentMapper.class);
        Stu s=new Stu();
        s.setId(103);
        s.setName("jerry");
        s.setAge(19);
        mapper.insertStu(s);
        ss.commit();
        ss.close();
    }
}
```

【程序解析】 session 对象的 getMapper 方法，是完成动态代理实现 DAO 接口的关键方法，该方法通过挂载一个接口，依靠 MyBatis 框架的代理机制，自动实现接口中的方法并得到返回值。

本章小结

本章讲解了 MyBatis 中动态代理实现 DAO 接口的基本原理和具体的开发方式。读者通过与非动态代理进行对比，感受使用动态代理的便捷性，从而加深对 MyBatis 框架原理的理解。

习题

简述 MyBatis 框架中使用动态代理实现 DAO 接口的优势。

上机指导

使用 MyBatis 框架动态代理实现 DAO 接口，完成查询数据库中 users 表的所有数据并遍历。

10 第10章　MyBatis与Spring的整合

学习目标

- 了解 MyBatis 与 Spring 框架整合的优势
- 掌握 MyBatis 与 Spring 框架整合的开发方式

MyBatis 和 Spring 两大框架的组合已经成了 Java 互联网技术主流的框架组合，它们经受住了大数据量和大批量请求的考验，在互联网系统中得到了广泛的应用。MyBatis 实现了对数据进行简单、方便、快捷的持久化访问，Spring 框架具有 IoC 和 AOP 的技术功能，两者都对项目开发以及代码的规范起到了非常重要的推进作用。这两个框架是可以同时使用的，本章将介绍 MyBatis 和 Spring 框架的整合使用。

10.1　MyBatis 与 Spring 框架整合的优势

MyBatis 与 Spring 框架整合的优势

目前不少的 Java 互联网项目，都是使用 Spring + MyBatis 搭建平台的。使用 Spring IoC 可以有效地管理后台的 Java 资源，通过配置达到即插即用的可拔插功能；通过 Spring AOP 框架，可以实现面向切面的代码开发，很大程度上消除了开发中的冗余代码。而 MyBatis 框架对数据库访问具有高灵活、可配置、可优化 SQL 等特性，两者结合使用完全可以构建高性能的大型网站。

MyBatis 与 Spring 的整合会将 MyBatis 代码无缝地整合到 Spring 中。它将允许 MyBatis 参与到 Spring 的事务管理中，通过 Spring 中内置的功能组件，更好地完成对数据库的访问操作。

10.2　MyBatis 与 Spring 框架整合案例

前面我们已经对 MyBatis 和 Spring 框架分别进行了介绍，这是进行框架整合的基础。框架的整合并不烦琐，只要根据一定的步骤，就能轻松地完成整合，在项目开发中这是个一劳永逸的过程，整合后程序员可以将自身精力更多地放到业务分析和代码设计中去。下面通过实际案例来介绍两个框架整合的具体方式。

【实例 10.1】使用框架整合对 stus 表添加记录。

（1）在 Eclipse 中新建 Web 项目，项目名设置为 MyBatis_Spring_Ch10。项目的组织结构如图 10-1 所示。

实例 10.1

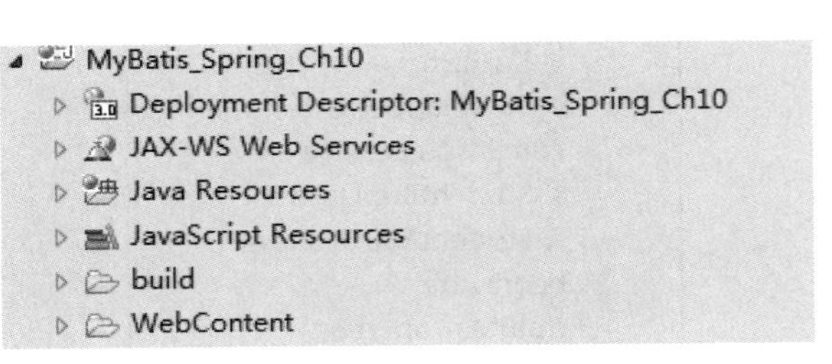

图 10-1　案例项目 MyBatis_Spring_Ch10 组织结构

（2）在项目的 WEB-INF/lib 目录下导入项目所需要的 JAR 包，包括 MyBatis 和 Spring 框架各自支持的 JAR 包。开发工具会自动为这些 JAR 包构建路径。导入的 JAR 包组织结构如图 10-2 所示。

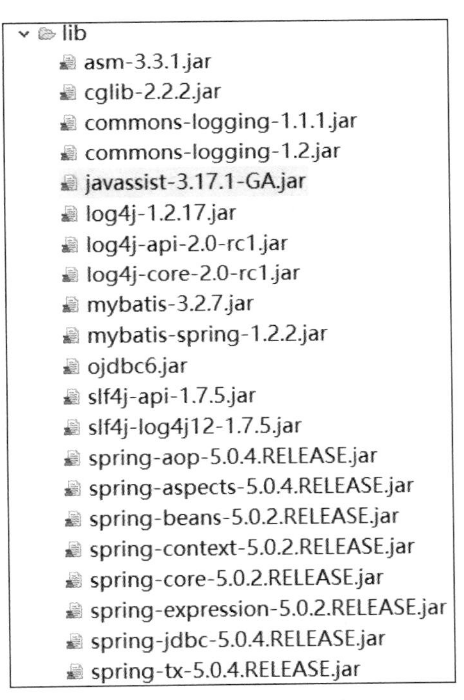

图 10-2　导入 JAR 包组织结构

这里并没有选用 Spring 框架下所有的 JAR 包，只选取其完成基本功能所必需的支持包，以及对 JDBC、切面和事务操作的支持 JAR 包。mybatis-spring-1.2.2.jar 支持包是 MyBatis 与 Spring 框架关联使用时，提供组件支持的 JAR 包，单独使用两个框架时不会用到该文件。

（3）在 Oracle 数据库中创建一个简单的 stus 表。代码如下。

```
create table stus(id number(5),name varchar2(50),age number(3));
```

（4）根据 MVC 开发思想，在 src 目录下新建代码结构，如图 10-3 所示。

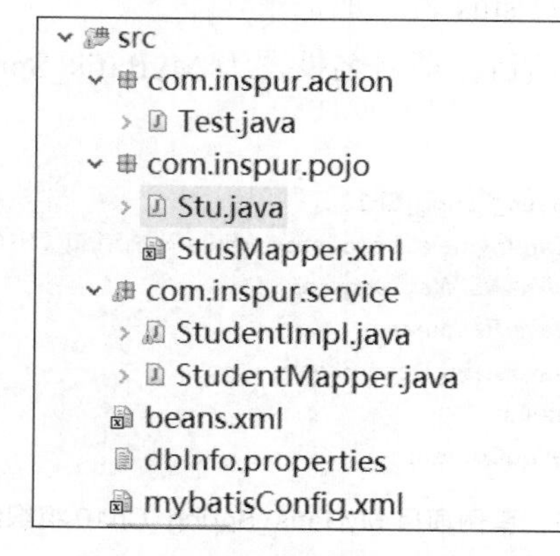

图 10-3　整合后的代码组织结构

com.inspur.action 包中的 Test 类作为测试的控制类；com.inspur.pojo 包中的 Stu 类作为实体类；StusMapper.xml 是 MyBatis 框架的映射文件；com.inspur.service 包下是 StudentMapper 接口及其实现类，本章我们采用 MyBatis 原生的 CRUD 操作；beans.xml 文件用来作为 Spring 框架的配置文件；dbInfo.properties 文件用来存放数据库的基本参数信息；mybatisConfig.xml 文件作为 MyBatis 框架的环境配置文件。

（5）完成 MyBatis 框架的相关配置。

```
dbInfo.properties
#Oracle
driver=oracle.jdbc.driver.OracleDriver
url=jdbc:oracle:thin:@127.0.0.1:1521:ORCL
userName=system
password=oracle
```

将数据库的基本参数信息配置到独立的属性文件中，从而更好地降低耦合度。其他文件可以通过调用该属性文件获取数据。

```
mybatisConfig.xml
<?xml version="1.0" encoding="UTF-8" ?>
<!DOCTYPE configuration
 PUBLIC "-//mybatis.org//DTD Config 3.0//EN"
 "http://mybatis.org/dtd/mybatis-3-config.dtd">
<configuration>
    <settings>
        <setting name="logImpl" value="STDOUT_LOGGING"></setting>
    </settings>
    <mappers>
        <mapper resource="com/inspur/pojo/StusMapper.xml"/>
    </mappers>
</configuration>
```

在 Spring 框架与 MyBatis 框架整合后，原本在 MyBatis 框架中配置的 Oracle 数据库信息，交给 Spring 框架统一管理。MyBatis 的配置文件得到了最大程度的简化。

```
StusMapper.xml
<?xml version="1.0" encoding="UTF-8"?>
<!DOCTYPE mapper
 PUBLIC "-//mybatis.org//DTD Mapper 3.0//EN"
 "http://mybatis.org/dtd/mybatis-3-mapper.dtd">
<mapper namespace="stu">
  <insert id="insertStu" parameterType="com.inspur.pojo.Stu">
      insert into stus values(#{id},#{name},#{age})
  </insert>
</mapper>
```

由于不使用 MyBatis 的动态代理，这里 mapper 的 namespace 属性用简单的字符串“stu”作为值即可。

（6）完成 Spring 框架相关的配置。

```
beans.xml
<?xml version="1.0" encoding="UTF-8"?>
<beans xmlns="http://www.springframework.org/schema/beans"
xmlns:xsi="http://www.w3.org/2001/XMLSchema-instance"
xmlns:context="http://www.springframework.org/schema/context"
xmlns:aop="http://www.springframework.org/schema/aop"
xmlns:tx="http://www.springframework.org/schema/tx"
xsi:schemaLocation="http://www.springframework.org/schema/beans
    http://www.springframework.org/schema/beans/spring-beans.xsd
    http://www.springframework.org/schema/context
    http://www.springframework.org/schema/context/spring-context.xsd
    http://www.springframework.org/schema/aop
    http://www.springframework.org/schema/aop/spring-aop.xsd
    http://www.springframework.org/schema/tx
    http://www.springframework.org/schema/tx/spring-tx.xsd">
    <!--开启注解扫描-->
    <context:component-scan base-package="com.inspur.*"></context:component-scan>
    <!-- 事务注解模式驱动-->
     <tx:annotation-driven transaction-manager="tm" />
     <bean id="tm" class="org.springframework.jdbc.datasource.DataSource
         TransactionManager">
         <property name="dataSource" ref="ds"></property>
     </bean>
     <!-- 在 Spring 的配置文件中引入 src 下的 properties 文件 -->
     <context:property-placeholder location="classpath:dbInfo.properties"/>
     <!--配置数据源-->
     <bean id="ds" class="org.apache.ibatis.datasource.pooled.PooledDataSource">
        <property name="driver" value="${driver}"></property>
        <property name="url" value="${url}"></property>
        <property name="username" value="${userName}"></property>
        <property name="password" value="${password}"></property>
        <!-- 配置最大活跃连接数和空余连接数 -->
        <property name="poolMaximumActiveConnections" value="3"></property>
        <property name="poolMaximumIdleConnections" value="1"></property>

     </bean>
     <!-- 关联 MyBatis 的配置，获取 session 工厂实例-->
```

```
        <import resource="classpath:mybatisConfig.xml"/>
        <bean id="ssf" class="org.mybatis.spring.SqlSessionFactoryBean">
            <property name="configLocation" value="classpath:mybatisConfig.xml">
            </property>
            <property name="dataSource" ref="ds"></property>
        </bean>
    </beans>
```

将 MyBatis 数据源的配置放到 Spring 框架下统一管理，有助于框架整合后的运行效率。使用 MyBatis 和 Spring 整合 JAR 包中的 SqlSessionFactoryBean 类，可以获取一个连接工厂的对象（ssf），方便了后台数据访问时的代码开发。

（7）完成后台代码的设计。

```
Stu.java
package com.inspur.pojo;
import org.springframework.context.annotation.Scope;
import org.springframework.stereotype.Component;
@Component("stu")
@Scope("prototype")
public class Stu {
    private int id;
    private String name;
    private int age;
    public int getId() {
        return id;
    }
    public void setId(int id) {
        this.id = id;
    }
    public String getName() {
        return name;
    }
    public void setName(String name) {
        this.name = name;
    }
    public int getAge() {
        return age;
    }
    public void setAge(int age) {
        this.age = age;
    }
}
```

通过 Spring 注解创建一个 Stu 实体对象。

```
StudentMapper.java
package com.inspur.service;
import com.inspur.pojo.Stu;
public interface StudentMapper {
    void insertStu(Stu s);
}
```

接口中定义了插入一条学生记录的方法。

```
StudentImpl.java
package com.inspur.service;
```

```
import javax.annotation.Resource;
import org.apache.ibatis.session.SqlSession;
import org.apache.ibatis.session.SqlSessionFactory;
import org.springframework.context.annotation.Scope;
import org.springframework.stereotype.Component;
import org.springframework.transaction.annotation.EnableTransactionManagement;
import org.springframework.transaction.annotation.Transactional;
import com.inspur.pojo.Stu;
@Component("si")
@Transactional
@Scope("prototype")
@EnableTransactionManagement
public class StudentImpl implements StudentMapper {
    @Resource(name="ssf")
    private SqlSessionFactory ssf;
    public void insertStu(Stu s) {
        SqlSession ss=ssf.openSession();
        int i=ss.insert("stu.insertStu", s);
        ss.close();
    }
}
```

这是StudentMapper类的实现类。为了更清晰地看到对象使用的过程，这里没有使用MyBatis的动态代理。使用 Spring 事务注解完成事务操作时，这里应对该类添加@EnableTransaction-Management注解以启用事务管理。ssf对象是在beans.xml中配置的数据库连接工厂对象。

```
Test.java 文件
package com.inspur.action;
import java.io.IOException;
import java.io.InputStream;
import org.apache.ibatis.io.Resources;
import org.apache.ibatis.session.SqlSession;
import org.apache.ibatis.session.SqlSessionFactory;
import org.apache.ibatis.session.SqlSessionFactoryBuilder;
import org.springframework.context.ApplicationContext;
import org.springframework.context.support.ClassPathXmlApplicationContext;
import com.inspur.pojo.Stu;
import com.inspur.service.StudentImpl;
import com.inspur.service.StudentMapper;
public class Test {
    public static void main(String[] args) {
        ApplicationContext ac=ac=new ClassPathXmlApplicationContext("beans.xml");
        Stu s=(Stu) ac.getBean("stu");
        StudentImpl si=(StudentImpl) ac.getBean("si");
        s.setId(101);
        s.setName("tom");
        s.setAge(20);
        si.insertStu(s);
    }
}
```

运行测试后，从Oracle的开发工具中查询到stus表中已经插入了一条记录，如图10-4所示。

ID	NAME	AGE
101	tom	20

图10-4 案例项目运行结果

本章小结

本章讲解了 MyBatis 框架与 Spring 框架整合使用的优势和必要性，以及两个框架整合的具体流程。特别是涉及了事务的操作在 Spring 中的配置，以及数据源在 Spring 中的配置和使用。在实际的项目开发中，也可能会用到数据连接池等第三方插件作为数据源，读者可以另行查询相关资料，以扩展自己的知识面。读者通过本章的学习会发现，整合过程并非简单的拼凑，而是涉及了一些配置文件的相互引用关联。

习题

1. 简述 MyBatis 框架和 Spring 框架整合使用的优势。
2. 在 MyBatis 框架和 Spring 框架的整合过程中，使用事务需注意哪些细节？

上机指导

整合 MyBatis 框架和 Spring 框架，以便进行数据库中表信息的查询。

第11章 Spring MVC入门

学习目标

- 了解 Spring MVC 框架及其核心功能组件
- 掌握 Spring MVC 框架的请求处理流程
- 了解 Spring MVC 框架的优势
- 理解 Spring MVC 框架入门案例

Spring MVC 是一种轻量级的、基于 MVC 的 Web 应用框架，是目前较好的实现 MVC 设计模式的框架之一，在实际开发中也常被叫作 Spring MVC 框架。顾名思义，Spring MVC 是 Spring 框架的一个分支产品，它以 Spring IoC 容器为基础，并利用容器的特性来简化它的配置。作为 Spring 的一个 Web 组件，它为构建稳健的 Web 应用提供了丰富的功能。

11.1 Spring MVC 框架概述

Spring MVC 框架概述

11.1.1 Spring MVC 框架的核心功能介绍

通过之前的讲解我们可以知道，MVC 是模型（Model）—视图（View）—控制器（Controller）的缩写。通过代码的分层，既能保证代码的整洁性，又能保证代码具有较高的可维护性，同时还有利于团队开发，提高整体开发效率。各层次之间的关系如图 11-1 所示。

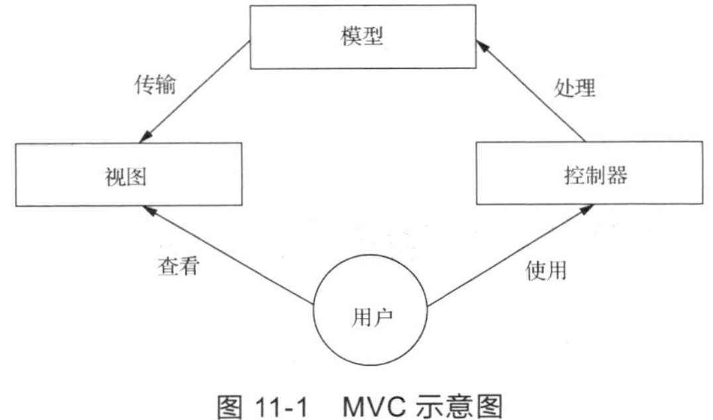

图 11-1 MVC 示意图

Spring MVC 是一个基于 MVC 开发思想的具体实现框架。在该框架中，用户的每一个请求都声明了一个需要执行的动作。其实现方式是通过将每个请求路径（URI）映射到一个可执行的方法来实现。同时，也将请求中的参数数据映射到对应方法的参数变量中。

Spring MVC 提供了数据绑定机制。通过该机制，在遵循一定的代码规则的基础上，可以从视图层（View）的用户请求中提取数据，并将数据转换为控制层（Controller）中预定义的数据格式，最后映射到一个模型类（Model），进而创建一个对象。

Spring MVC 通过其视图处理机制，根据控制层代码中设置的视图路径，自动实现了请求路径的转换，并可以通过参数配置实现不同的路径转换方式。而且该视图处理机制并没有规定必须使用哪种视图技术，可以是 JSP，也可以是 Velocity 模板、Tiles 及 XSLT 等。为视图开发提供了强大的支持。

除此之外，Spring MVC 还具有其他一些功能特征。

（1）Spring MVC 框架与 Spring 框架其他组件模块的无缝集成。Spring 是一个一站式的框架，提供了表现层（Spring MVC）到业务层（Spring）再到数据层（Spring Data）的全套解决方案。Spring 的两大核心 IoC（控制反转）和 AOP（面向切面编程）更是给程序解耦提供了支持，目前已在项目开发中得到了广泛的使用。Spring MVC 作为 Spring 的一个 Web 组件，可以实现天然的集成使用，给开发带来极大的方便。

（2）Spring MVC 框架提供了强大而直接的配置方式：将框架类和应用程序类作为 JavaBean 配置，支持跨多个 Context 的引用，例如，在 Web 控制器中对业务对象和验证器（Validator）的引用。

（3）使用 Spring MVC 框架能简单地进行 Web 层的单元测试，提高开发效率。

（4）使用 Spring MVC 框架便于与其他视图技术集成，如 Velocity、FreeMarker 等，因为模型数据放在一个 Model 里（Map 数据结构实现，因此很容易被其他框架使用）。

（5）Spring MVC 框架提供了非常灵活的数据验证、格式化和数据绑定机制。

（6）Spring MVC 框架提供一套强大的 JSP 标签库来简化 JSP 开发。

（7）支持灵活的本地化、主题等解析。

（8）Spring MVC 框架提供了统一异常处理机制，使得异常处理更加简单。

11.1.2 Spring MVC 框架的核心组件构成

Spring MVC 实现了 Web 层的开发规范，其底层依赖于一系列的功能组件，配合完成整个功能。其基本的运行流程和使用到的核心组件如图 11-2 所示。

在 Spring MVC 框架工作的过程中，主要的几个核心组件内容，通过相互之间的协调运行，共同完成框架的核心功能，下一节将详细介绍这几个核心组件。

11.2 Spring MVC 框架的工作流程

Spring MVC 框架的工作流程

前面我们介绍了 Spring MVC 框架的核心组件及其功能，本节将介绍框架对请求的处理流程及期间使用的主要类和接口，以加深大家对框架的了解。

11.2.1 Spring MVC 框架的请求执行顺序

从用户的浏览器发出访问请求开始，到后台服务器经过一系列的操作，最终返回给用户视图界面，框架的核心架构和工作流程如图 11-2 所示。

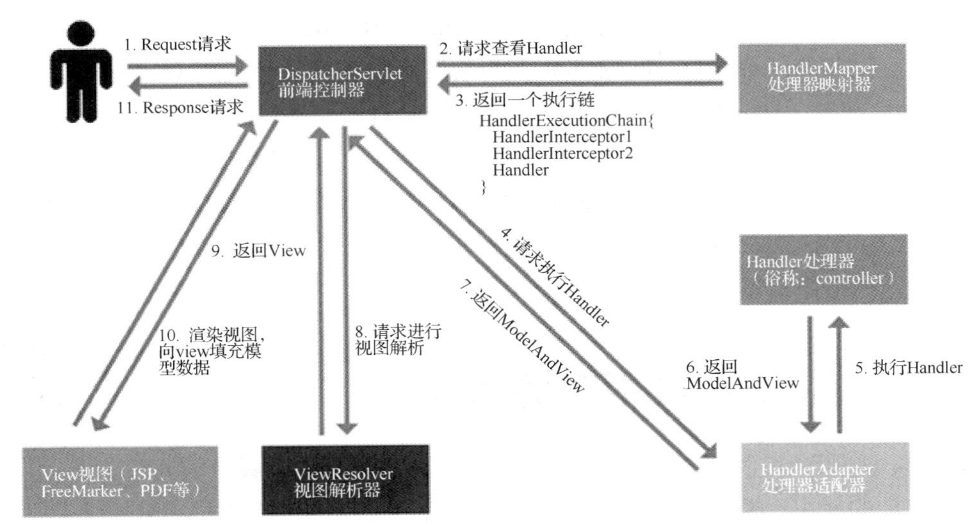

图 11-2 Spring MVC 核心架构和工作流程

（1）用户向服务端发送一次请求，在满足 web.xml 文件中前端控制器的匹配规则的前提下，这个请求在 Tomcat 的解析下，会先到前端控制器 DispatcherServlet（也叫中央控制器）。如我们发出*.action 的请求，则只对以“.action”结尾的请求路径进行处理。某些访问 JSP 的页面请求则不会被框架拦截处理。

（2）DispatcherServlet 作为框架的核心，通过处理器映射器（框架提供 SimpleUrl HandlerMapping 类，也可能是其他的映射器类），从映射文件中根据请求路径找到对应的处理器类，并得到该类的实例对象。

（3）DispatcherServlet 通过处理器适配器（框架提供的一个类），对处理类的实例对象进行适配器处理以及参数绑定，并将请求转移到处理类中进行业务处理。

（4）处理类通过实现框架提供的 Controller 接口，最终完成适配功能。并在这里完成具体的业务逻辑功能，将数据和视图信息封装到 ModelAndView 对象中，并返回给 DispatcherServlet 处理。

（5）DispatcherServlet 获取从处理器返回的 ModelAndView 对象后，交给视图解析器（框架中提供的功能类）解析，并显示在相应的视图中。

11.2.2 Spring MVC 框架的核心接口

Spring MVC 框架实现处理请求的整个过程，需要各个功能组件的协调运行，下面分析这个过程中主要用到的类或接口。

（1）前端控制器（DispatcherServlet）：它也称为中央控制器，是整个请求响应的控制中心，组件的调用都由它统一调度。配置在 web.xml 文件中，随服务器启动而实例化。

（2）简单处理器映射器（SimpleUrlHandlerMapping）：根据用户访问的 URL 映射到对应的后端处理器 Handler，并将该处理器的实例返回给前端控制器。在 Spring MVC 框架中还提供了其他的处理器映射器，同样可以完成此功能。

（3）处理器适配器（HandlerAdapter）：常用的有 SimpleControllerHandlerAdapter，该功能组件在框架中默认运行。它用于封装参数数据，绑定视图等。

（4）视图解析器（ViewResolver）：常用的有 InternalResourceViewResolver，在框架中该功能组件默认运行。它用于将 ModelAndView 逻辑视图解析为具体的视图（如 JSP）。

（5）后端处理器（Handler）：即由程序员编写的处理类，对用户具体请求进行处理，该类必须实现框架提供的 Controller 接口，才能作为处理器完成组件功能。

（6）处理器中使用的类（ModelAndView）：用于封装数据和视图信息，并返回给前端控制器。该类实现了对原生 Servlet 中 Request 对象的封装，因此可以作为域对象封装数据，并在视图中用 EL 表达式取出数据。

11.3 Spring MVC 框架的优势

Spring MVC 框架的优势

MVC 即模型-视图-控制器，是 20 世纪 80 年代出现的一种软件设计模式，现在已得到广泛的使用。模型是应用程序的主体部分，它表示业务数据或者业务逻辑；视图是应用程序中与用户界面相关的部分，是用户看到并与之交互的界面；控制器的工作就是根据用户的输入，控制用户界面数据显示和更新 Model 对象状态。MVC 模式的出现不仅实现了功能模块和显示模块的分离，同时它还提高了应用系统的可维护性、可扩展性、可移植性和组件的可复用性。

SUN 公司在 JSP 出现的早期就制定了两种规范：Model1 和 Model2。Model1 对 MVC 模式的实现并不太令人满意，而 Model2 虽然在一定程度上实现了 MVC 的设计模式，但也有诸多弊端，例如容易使系统出现多个 Controller，且对页面导航的处理比较复杂。此后，又出现了 Struts、Struts2 等框架来实现 MVC 设计模式，Spring MVC 框架也应运而生了，且对比其他框架有着显著的优势。

（1）Spring MVC 的框架机制是通过 Servlet 技术的方式进行拦截，在第一次请求发送时初始化，并随着容器关闭而销毁。

（2）Spring MVC 具有独立的 AOP 拦截机制。它是方法级别的拦截，一个请求对应着一个处理器中的方法，请求参数会封装到方法参数中。

（3）Spring MVC 对注解的支持非常优秀，在配置了基本的参数之后，再编写处理器类和方法时，只需加上注解即可，无须频繁修改配置文件。

（4）Spring MVC 和 Spring 同根同源，整合时无须进行配置。

11.4 Spring MVC 实现模拟登录

Spring MVC 实现模拟登录

11.4.1 Spring MVC 开发环境

进行 Spring MVC 开发需要进行一些环境配置。

Spring MVC 是基于 MVC 的 Web 应用框架，其应用与 Web 服务器密切相关。Tomcat 因其技术先进、性能稳定，而且免费，深受 Java 爱好者的喜爱并得到了部分软件开发商的认可，成为目前比较流行的 Web 应用服务器。本章我们依然采用 Tomcat 服务器进行讲解。Tomcat 可以从官网上下载，本章使用的是从 Apache 官网下载的 Tomcat8 版本、Spring MVC 支持包，如图 11-3 所示。Spring MVC 框架功能的实现依赖于 JAR 包的支持，我们通过前面下载 Spring 框架的完整产品，可以找到多个 JAR 包来为不同的功能实现提供支持。这里选用最基础也最为常用的 6 个 JAR 包，以及通用的日志 JAR 包，共同完成本案例的实现。

commons-logging-1.1.1.jar
spring-beans-5.0.2.RELEASE.jar
spring-context-5.0.2.RELEASE.jar
spring-core-5.0.2.RELEASE.jar
spring-expression-5.0.2.RELEASE.jar
spring-web-5.0.4.RELEASE.jar
spring-webmvc-5.0.4.RELEASE.jar

图 11-3　Spring MVC 导入 JAR 包示意图

11.4.2　使用 Spring MVC 完成登录验证

本节将通过一个具体的登录验证案例，来介绍框架的具体开发方式和工作流程。

【实例 11.1】使用 Spring MVC 框架进行用户登录验证，这里暂时不使用数据库操作。

具体开发方式如下（代码详见 Springmvc_Ch11\WEB-INF\web.xml；Springmvc_Ch11\WEB-INF\springmvc-servlet.xml；Springmvc_Ch11\login.jsp；Springmvc_Ch11\success.jsp；Springmvc_Ch11\src\com\inspur\action\LoginController.java）。

1. 新建 Web 项目

打开 Eclipse，建立 Dynamic Web Project 项目，因这里需要使用浏览器与 Tomcat 服务器的交互，必须建立 Web 项目。在创建过程中，需要关联 Tomcat 服务器，通过“New Runtime”选择 Tomcat 服务器。选择 Tomcat 服务器后，进行“next”操作，进入 Java 的源代码文件夹配置界面，这里不需要做任何修改，继续进行“next”操作，进入 Web Module 的配置界面，这里选中创建 web.xml 文件（它是 Web 开发中非常重要的一个配置文件），如图 11-4 所示。

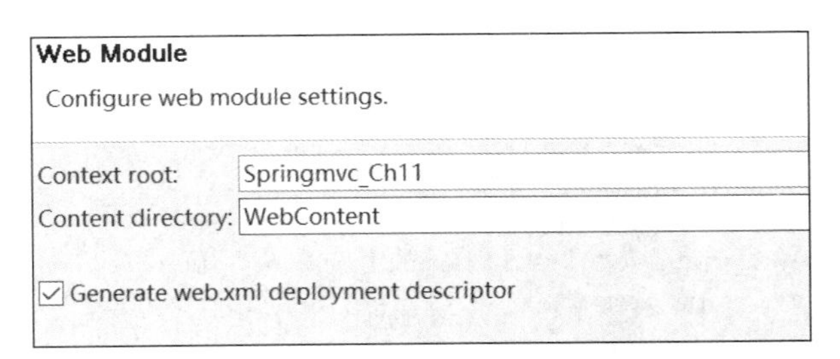

图 11-4　选中创建 web.xml 示意图

2. 导入支持的 JAR 包

导入路径。复制图 11-3 所示的 7 个支持包，粘贴到路径 WebContent/WEB-INF/lib 目录下，在 Eclipse 下可以自动构建路径。目录结构如图 11-5 所示。

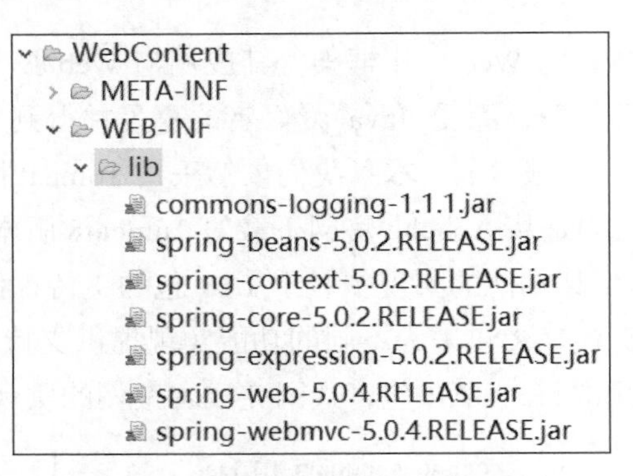

图 11-5 入门案例项目需导入的 JAR 包

这里对引入的 JAR 包进行简单介绍。

（1）commons-logging：提供了一个日志接口，Spring 的内部实现中使用了该 JAR 包下的接口，这是 Spring 开发必须引入的一个支持包。

（2）spring-beans：该支持包是 Spring IoC 开发必不可缺的支持包。

（3）spring-context：这个 JAR 文件为 Spring 核心提供了大量的扩展功能，在实际开发中它是必须引入的一个支持包。

（4）spring-core：这个 JAR 文件包含 Spring 框架基本的核心工具类，Spring 其他组件都要用到这个包里的类。

（5）spring-expression：Spring 表达式语言的支持包。

（6）spring-web：Spring MVC 支持 Web 端应用部署架构的支持包。

（7）spring-webmvc：REST Web 服务和 Web 应用的试图控制器支持包。

3. 配置 Spring MVC 的前端控制器

Spring MVC 框架通过前端控制器对符合匹配规则的客户端请求进行统一的拦截处理。这些匹配规则放置在 Spring MVC 的配置文件中，框架通过默认的路径去加载这些文件（也可通过某些操作来修改其默认路径，这里暂时不介绍）。

配置文件路径：在 WebRoot/WEB-INF/web.xml 中进行配置。

配置内容：由于 Spring MVC 框架的前端控制器是一个 Servlet，该 Servlet 在前面的支持 JAR 包中存在。其配置方式与 Servlet 的配置一致，具体代码如下。

```
web.xml
<?xml version="1.0" encoding="UTF-8"?>
<web-app xmlns:xsi="http://www.w3.org/2001/XMLSchema-instance" xmlns="http://
java.sun.com/xml/ns/javaee" xsi:schemaLocation="http://java.sun.com/xml/ns/
javaee http://java.sun.com/xml/ns/javaee/web-app_3_0.xsd" id="WebApp_ID"
version="3.0">
  <display-name>Springmvc_Ch11</display-name>
  <welcome-file-list>
    <welcome-file>index.html</welcome-file>
    <welcome-file>index.htm</welcome-file>
    <welcome-file>index.jsp</welcome-file>
    <welcome-file>default.html</welcome-file>
    <welcome-file>default.htm</welcome-file>
```

```
        <welcome-file>default.jsp</welcome-file>
    </welcome-file-list>
    <!-- 配置 Spring MVC 的前端控制器 -->
    <servlet>
        <servlet-name>springmvc</servlet-name>
        <servlet-class>org.springframework.web.servlet.DispatcherServlet</servlet- class>
    </servlet>
    <servlet-mapping>
        <servlet-name>springmvc</servlet-name>
        <!-- 匹配所有以.action 结尾的请求 -->
        <url-pattern>*.action</url-pattern>
    </servlet-mapping>
</web-app>
```

4. 配置 Spring MVC 的核心映射文件 springmvc–servlet.xml

文件路径：在 WebContent/WEB-INF 目录下新建 springmvc-servlet.xml。文件名来源于 web.xml 配置的前端控制器的名称与后缀“-servlet”，前端控制器在加载过程中会自动在该路径下寻找该映射文件。因此该文件的路径和文件名不能随意更改。如要更改，则需在 web.xml 中配置前端控制器时，修改其默认的映射文件路径和文件名，这里暂时不做讲解。

配置内容如下。

（1）xml 文件中的约束可以通过下载框架中的参考文档查找，也可以从本文中复制。

（2）配置访问路径与后台处理器之间的映射。该配置需要框架提供的映射类 SimpleUrl-HandlerMapping 来完成请求路径的跳转。

配置代码如下。

```
springmvc-servlet.xml
<?xmlversion="1.0"encoding="UTF-8"?>
<beansxmlns="http://www.springframework.org/schema/beans"
xmlns:xsi="http://www.w3.org/2001/XMLSchema-instance"
xsi:schemaLocation="http://www.springframework.org/schema/beans
http://www.springframework.org/schema/beans/spring-beans.xsd">
<!-- 配置映射关系 -->
    <bean id="lc" class="control.LoginController"></bean>
    <bean class="org.springframework.web.servlet.handler.SimpleUrlHandlerMapping">
        <property name="mappings">
            <props>
                <!-- 具体映射关系：url（站内路径）和处理器（handler）的映射 -->
                <prop key="/lc.action">lc</prop>
            </props>
        </property>
    </bean>
</beans>
```

5. 新建登录页面 login.jsp 和登录成功后的页面 success.jsp

文件路径：在 WebRoot 目录下新建 jsp 页面（login.jsp 和 success.jsp）。

具体代码如下。

```
login.jsp
<form action="/Springmvc_Ch11/lc.action"method="post">
```

```
请输入用户名:<inputtype="text"name="userName"/><br/>
请输入密码:<inputtype="password"name="password"/><br/>
<input type="submit"value="点击登录 "/>
</form>
success.jsp
<body>
<h2>登录成功页面</h2>
<!-- 使用 EL 表达式取出 mav 中的数据 -->
    ${info}
</body>
```

6. 新建后台登录处理器类 LoginController.java

该类的名称必须与映射文件中的配置保持一致。

文件路径：在 src/control 下新建普通的 class 文件 LoginController.java。

文件内容：进行登录的模拟验证处理，具体代码如下。

```
LoginController.java
package com.inspur.action;
import javax.servlet.http.HttpServletRequest;
import javax.servlet.http.HttpServletResponse;
import org.springframework.web.servlet.ModelAndView;
import org.springframework.web.servlet.mvc.Controller;
public class LoginController implements Controller{
    @Override
    public ModelAndView handleRequest(HttpServletRequest request,
        HttpServletResponse response) throws Exception {
        String userName=request.getParameter("userName");
        String password=request.getParameter("password");
        ModelAndView mav=new ModelAndView();
        if("123".equals(password)){
            mav.addObject("info", "欢迎你"+userName);
            mav.setViewName("success.jsp");
        }
        return mav;
    }
}
```

【程序解析】

在 Spring MVC 框架中进行业务处理的类（控制器），需要框架提供的 Controller 接口，才能实现映射关系。进而程序才能通过访问路径进入该类的 handleRequest 方法中运行。

ModelAndView 类是 Spring MVC 框架提供的用于数据封装和视图跳转的类。通过该类的 addObject 方法可以进行数据的封装，其功能类似 Servlet 中的 request 对象。通过该类的 setViewName 方法可以进行视图路径的跳转，当对象返回前端控制器时，可能引发视图转换。

7. 登录

启动服务器，在浏览器中访问 login.jsp 页面，如图 11-6 所示。输入任意用户名，密码输入“123”，单击“点击登录”按钮，顺利进入登录成功页面。显示界面如图 11-7 所示。

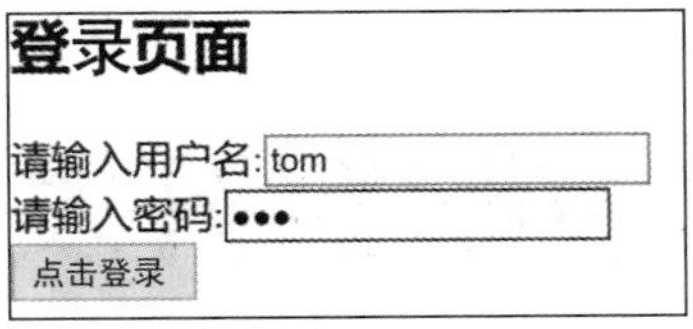

登录页面

请输入用户名: tom
请输入密码: •••
点击登录

图 11-6　案例项目登录界面图

图 11-7　案例项目登录成功界面图

本章小结

本章介绍了 Spring MVC 框架的概念、特征及核心功能，读者需要掌握该框架在 Web 应用中的作用及其工作原理。了解以上内容后，再通过入门案例掌握 Spring MVC 开发所需要的开发环境以及基本的开发步骤，理解利用框架开发的优点，体会框架开发的便利之处，同时学会使用框架提供的其他功能组件进行快速开发。

习题

1. 简述 Spring MVC 框架及其工作原理。
2. 简述 Spring MVC 框架的核心组件。

上机指导

开发一个 Web 工程，并引入 Spring MVC 框架，完成模拟用户注册功能。

12

第12章 注解式控制器开发

学习目标

- 掌握注解式控制器的概念
- 掌握使用注解完成 Spring MVC 的控制器基本功能的方法
- 掌握使用注解完成 Spring MVC 的参数绑定的方法
- 掌握使用注解完成 Spring MVC 参数绑定时的类型转换的方法
- 掌握常见的 Spring MVC 注解及其属性配置
- 了解 REST 的概念

12.1 注解式控制器的概念

注解式控制器的概念

1. Spring 2.5 注解

Spring 2.5 之前，都是通过 Controller 接口或其实现来定义处理器类的。

Spring 2.5 引入注解式处理器支持，通过@Controller 和@RequestMapping 注解定义处理器类，并且提供了一组强大的注解。

@Controller：用于标识处理器类。

@RequestMapping：用于从客户端到控制器的地址映射。

@RequestParam：用于客户端参数注入控制器的数据绑定。

@ModelAttribute：用于请求参数到命令对象的绑定。

@SessionAttributes：该注解可将 request 中存在的值共享到session 中。

@InitBinder：自定义数据绑定注册支持，用于将请求参数转换到命令对象属性的对应类型；用于在@Controller 中标注方法，表示为当前控制器注册一个属性编辑器或者其他，只对当前的 Controller 有效。

需要通过处理器映射 DefaultAnnotationHandlerMapping 和处理器适配器 AnnotationMethodHandlerAdapter 来开启支持@Controller 和@RequestMapping 注解的处理器。

2. Spring 3.0 注解

Spring 3.0 版本以上引入了对 RESTful 架构风格的支持（通过@PathVariable 注解和一些其他特性来支持），且又引入了更多的注解支持。下面对注解功能进行简要说明。

@CookieValue：Cookie 数据到处理器功能处理方法的方法参数上的绑定。

@RequestHeader：请求头（Header）数据到处理器功能处理方法的方法参数上的绑定。

@RequestBody：请求的 Body 体的绑定（通过 HttpMessageConverter 进行类型转换）。

@ResponseBody：处理器功能处理方法的返回值将作为响应体（通过 HttpMessageConverter 进行类型转换）。

@ResponseStatus：定义处理器功能处理方法/异常处理器返回的状态码和原因。

@ExceptionHandler：注解式声明异常处理器。

@PathVariable：请求 URI 中的模板变量部分到处理器功能处理方法的方法参数上的绑定，从而支持 RESTful 架构风格的 URI。

自 Spring 3.0 版本以上用新的 HandlerMapping 和 HandlerAdapter 来支持 @Controller 和 @RequestMapping 注解处理器，组合使用处理器映射 RequestMappingHandlerMapping 和处理器适配器 RequestMappingHandlerAdapter 来代替 Spring 2.5 版本支持的处理器映射 DefaultAnnotationHandlerMapping 和处理器适配器 AnnotationMethodHandlerAdapter。

12.2　Spring MVC 实现控制器基本功能

Spring MVC 实现控制器基本功能

Spring 2.5 版本之前，都是通过 Controller 接口或其实现来定义处理器类的。随着 Spring 框架的迅速普及，单纯使用该接口来实现控制器，使得在用代码实现时，一定程度上降低了开发效率。为此 Spring 后续版本提供了强大的注解功能，不仅可以完成 Web 层最基本的控制器功能，更集成了许多简单实用的小功能，极大提高了程序员的开发效率。

本节将通过一个登录验证的案例，实用 Spring 注解的方式来实现控制器，帮助读者体会注解在 Spring MVC 开发中的特点。

【实例 12.1】使用 Spring MVC 框架进行用户登录验证，这里暂时不使用数据库操作。具体开发方式如下（代码详见 Springmvc_Ch12\WEB-INF\web.xml；Springmvc_Ch12\WEB-INF\springmvc-servlet.xml；Springmvc_Ch12\login.jsp；Springmvc_Ch12\success.jsp；Springmvc_Ch12\src\ com\inspur\action\LoginController.java）。

由于大部分代码与第 11 章中的案例相似，这里只截取使用注解完成控制器的部分。

（1）配置 Spring MVC 的核心映射文件 springmvc-servlet.xml。

```
<?xml version="1.0" encoding="UTF-8"?>
<!--增加 xmlns:context 命名空间及其约束：注解的相关配置  -->
<beans xmlns="http://www.springframework.org/schema/beans"
        xmlns:xsi="http://www.w3.org/2001/XMLSchema-instance"
        xmlns:context="http://www.springframework.org/schema/context"
        xmlns:mvc="http://www.springframework.org/schema/mvc"
```

```
        xsi:schemaLocation="http://www.springframework.org/schema/beans
           http://www.springframework.org/schema/beans/spring-beans.xsd
           http://www.springframework.org/schema/context
           http://www.springframework.org/schema/context/spring-context.xsd
           http://www.springframework.org/schema/mvc
           http://www.springframework.org/schema/mvc/spring-mvc.xsd">
    <!-- 开启注解扫描（默认状态下，不开启注解扫描）-->
    <context:component-scan base-package="com.inspur.action"></context:
component-scan>
</beans>
```

通过与第 11 章的案例的比较，发现这里在配置文件时，增加了开启注解扫描，同时引入了有关 context 的相关约束。而第 11 章中存在的关于登录类的 bean 实例，以及地址映射关系的配置，这里都不需要了，在代码中会通过注解来实现。

（2）新建后台登录处理器类 LoginController.java。具体代码如下。

```
LoginController.java
package com.inspur.action;
import javax.servlet.http.HttpServletRequest;
import javax.servlet.http.HttpServletResponse;
import org.springframework.stereotype.Controller;
import org.springframework.web.bind.annotation.RequestMapping;
import org.springframework.web.servlet.ModelAndView;
@Controller
public class LoginController{
    @RequestMapping("/loginCheck.action")
    public ModelAndView loginCheck(HttpServletRequest request,
        HttpServletResponse response) throws Exception {
        ModelAndView mav=new ModelAndView();
        String name=request.getParameter("userName");
        String password=request.getParameter("password");
        if("123".equals(password)){
            mav.addObject("info", "欢迎你"+name);
            mav.setViewName("success.jsp");
        }
        return mav;
    }
}
```

这里使用了 Spring MVC 完成控制器功能最核心的两个注解。

① @Controller 用于标记在一个类上，使用它标记的类就是一个 Spring MVC Controller 对象。分发处理器将会扫描使用了该注解的类的方法，对照第 11 章中通过接口完成控制器的操作，可以看到该注解还会对它标记的类进行实例化，并完成控制器的相关配置。但同时又解耦了控制器接口，使得程序员可以自由开发该类下的具体方法。

② @RequestMapping 是一个用来处理请求地址映射的注解，可用于类或方法上。用于类上，表示类中的所有响应请求的方法都是以该地址作为父路径。通过与第 11 章的比较可以看到，该注解替代了配置文件中的 URL 路径映射配置，通常使用该注解标记一个具体的方法，与前端 JSP 中某个访问路径匹配。

（3）新建后台登录界面 login.jsp。具体代码如下。

```
<body>
```

```
        <h3>登录页面</h3>
        <form action="/Springmvc_Ch12/loginCheck.action" method="post">
            请输入用户名：<input type="text" name="userName"/><br/>
            请输入密码：<input type="password" name="password"/><br/>
            <input type="submit" value="点击登录"/>
        </form>
    </body>
```

表单中的路径，通过 Web 服务器和框架，会与注解中的路径进行匹配，如果匹配成功，则进入相应的类或方法运行代码。这里的请求会进入控制器中的 loginCheck 方法。

（4）使用注解对请求方式进行匹配处理。在控制类中补充 test 方法代码如下。

```
@RequestMapping(value="/test.action",method={RequestMethod.GET})
public String test(){
    System.out.println("aaaaaaaaa");
    return "success.jsp";
}
```

使用注解有利于在控制类中进行具体业务方法的扩展，@RequestMapping 注解还可以通过对 method 属性的配置进行特殊的请求方式的匹配，如果匹配成功则进入具体方法。除了 value 和 method 属性，该注解还有其他一些属性配置，读者可以自己学习研究。

12.3　实现数据参数绑定

Spring MVC 框架为了提高开发者的开发效率，提供了很多的参数绑定功能，这些功能为开发者节约了大量的重复工作，下面介绍参数绑定。

实现数据参数绑定

（1）Spring MVC 默认支持的参数绑定。

在前面的用户登录案例中，当路径匹配成功后，程序进入具体的业务方法，此时进行登录模拟验证时，直接使用了 HttpServletRequest request 对象获取了页面参数值。这里的 request 对象就是框架默认封装好，供开发者直接使用的对象，除此之外，还有 HttpServletResponse 对象、HttpSession 对象也是比较常用的默认绑定参数对象。

（2）通过@RequestParam 注解，实现基本类型和字符串参数值的绑定。

Spring MVC 框架为方便实际开发时，大量页面参数值需要通过请求对象传递到后端，使用注解的形式直接将参数数据注入后端程序中，但必须按照一定的规则才能顺利完成绑定和注入。这里在登录页面中增加一个表单，代码如下。

```
<form action="/Springmvc_Ch12/loginCheck2.action" method="post">
    请输入用户名：<input type="text" name="name"/><br/>
    请输入密码：<input type="password" name="psd"/><br/>
    <input type="submit" value="点击登录"/>
</form>
```

在控制器中增加一个业务方法，代码如下。

```
@RequestMapping("/loginCheck2.action")
```

```
    public ModelAndView loginCheck2(
            @RequestParam("name") String userName,
            @RequestParam("psd") String password) throws Exception {
        ModelAndView mav=new ModelAndView();
        if("123".equals(password)){
            mav.addObject("info", "欢迎你"+userName);
            mav.setViewName("success.jsp");
        }
        return mav;
    }
```

【程序解析】Spring MVC 框架通过解析 form 表单中的 name 属性值，与@RequestParam 注解中的参数进行匹配，如果匹配成功，则将页面参数值直接赋给控制器中的变量。当使用直接注解修饰的变量名与注解中的参数一致时，可以直接省略注解。

（3）通过@RequestParam 注解，实现引用类型参数值的绑定。

在实际开发中，有时前端页面需要的数据非常杂乱，此时如果将这些数据封装成对象，则在后台接收数据时就相对方便了，节约了后台代码的书写量。Spring MVC 的参数映射注解同样提供了这样的功能。这里为了对象封装，先创建一个实体类 Users，代码如下。

```
    package com.inspur.entity;
    public class Users {
        private String name;
        private String password;
        private int age;
        public String getName() {
            return name;
        }
        public void setName(String name) {
            this.name = name;
        }
        public String getPassword() {
            return password;
        }
        public void setPassword(String password) {
            this.password = password;
        }
        public int getAge() {
            return age;
        }
        public void setAge(int age) {
            this.age = age;
        }
    }
```

在登录页面中增加一个表单，代码如下。

```
    <form action="/Springmvc_Ch12/loginCheck3.action" method="post">
        请输入用户名：<input type="text" name="name"/><br/>
        请输入密码：<input type="password" name=" password"/><br/>
        <input type="submit" value="点击登录"/>
    </form>
```

控制器中增加一个业务方法，代码如下。

```
    @RequestMapping("/loginCheck3.action")
    public ModelAndView loginCheck3(
            Users u) throws Exception {
        ModelAndView mav=new ModelAndView();
        if("123".equals(u.getPassword())){
            mav.addObject("info", "欢迎你"+u.getName());
            mav.setViewName("success.jsp");
        }
        return mav;
    }
```

【程序解析】Spring MVC 框架通过解析 form 表单中的 name 属性值，与控制器中的实体参数对象进行匹配，如果 name 属性值与实体参数对象的属性名一致，则将页面参数值直接封装成实体对象，并赋给控制器中的该实体对象的变量。这里并不需要写明具体注解。

12.4　实现参数类型转换

Spring MVC 框架的参数绑定功能，为提高开发效率提供了非常大的帮助，在实际开发中，页面参数都是以字符串的类型传递的，而在后台控制器中，需要的数据类型可能是多种多样的。有些类型必须由程序员进行手动转换，如字符串类型到 int 类型的转换。为了让程序员把更多的精力放到业务逻辑上，框架提供了一些类型的转换机制。

实现参数类型转换

1. 参数类型的自动转换

Spring MVC 框架可自动将页面输入的字符串类型数据，通过@RequestParam 注解转换为基本类型数据。如在登录页面中增加一个表单，代码如下。

```
    <form action="/Springmvc_Ch12/loginCheck4.action" method="post">
        请输入用户名：<input type="text" name="userId"/><br/>
        请输入密码：<input type="password" name="password"/><br/>
        <input type="submit" value="点击登录"/>
    </form>
```

控制器中增加一个业务方法，代码如下。

```
    @RequestMapping("/loginCheck4.action")
    public ModelAndView loginCheck4(
            @RequestParam("userId") int id,
            @RequestParam("password") String password) throws Exception {
        ModelAndView mav=new ModelAndView();
        if("123".equals(password)){
            mav.addObject("info", "欢迎你"+id);
            mav.setViewName("success.jsp");
        }
        return mav;
    }
```

【程序解析】Spring MVC 框架将页面中传入的字符串类型的 id，赋给控制器中的 int 类型的 id，整个过程不需要程序员处理，由框架自动完成。这里需要注意对前端数据在传输前进行校验，否则可能导致类型转换异常错误。字符串类型与数值类型的自动转换对数组也有效。

2. 字符串参数与其他非基本类型的转换

Spring MVC 框架除了提供对一些简单类型的自动转换之外，针对实际开发中的一些非基本类型，如日期类型等，也提供了与字符串类型之间的转换工具。这里通过增加一个表单，介绍该类型转换的方式。登录界面代码如下。

```
<form action="/Springmvc_Ch12/loginCheck5.action" method="post">
    请输入日期 A：<input type="text" name="dateTest"/><br/>
    请输入日期 B：<input type="text" name="dateTest"/><br/>
    <input type="submit" value="点击登录"/>
</form>
```

在控制器中增加一个业务方法，用来接收页面的日期数组，代码如下。

```
@RequestMapping("/loginCheck5.action")
public ModelAndView loginCheck5(
    @RequestParam("dateTest")Date[] date) throws Exception {
    ModelAndView mav=new ModelAndView();
    for(int i=0;i<date.length;i++){
        System.out.println(date[i]);
    }
        mav.setViewName("success.jsp");
        return mav;
    }
```

增加一个工具类文件 StringToDate.java，代码如下。

```
public class StringToDate implements Converter<String[], Date[]>{
    public Date[] convert(String[] returnBookDate) {
        Date[] dates=new Date[returnBookDate.length];
        SimpleDateFormat sdf=new SimpleDateFormat("yyyy-MM-dd");
        try {
            for(int i=0;i<returnBookDate.length;i++){
                Date date=sdf.parse(returnBookDate[i]);
                dates[i]=date;
            }

        } catch (ParseException e) {
            e.printStackTrace();
        }
        return dates;
    }
}
```

在 Spring MVC 的配置文件中，对该转换类进行配置，代码如下。

```
<?xml version="1.0" encoding="UTF-8"?>
<!--增加 xmlns:context 命名空间及其约束：注解的相关配置 -->
<beans xmlns="http://www.springframework.org/schema/beans"
    xmlns:xsi="http://www.w3.org/2001/XMLSchema-instance"
    xmlns:context="http://www.springframework.org/schema/context"
    xmlns:mvc="http://www.springframework.org/schema/mvc"
    xsi:schemaLocation="http://www.springframework.org/schema/beans
```

```
            http://www.springframework.org/schema/beans/spring-beans.xsd
            http://www.springframework.org/schema/context
            http://www.springframework.org/schema/context/spring-context.xsd
            http://www.springframework.org/schema/mvc
            http://www.springframework.org/schema/mvc/spring-mvc.xsd">
    <!-- 开启注解扫描（默认状态下，不开启注解扫描）-->
    <context:component-scan base-package="com.inspur.action"></context:component-
scan>
    <bean id="conversionService" class="org.springframework.format.support.Formatting
ConversionServiceFactoryBean">
        <property name="converters">
            <bean class="com.inspur.util.StringToDate"></bean>
        </property>
    </bean>
    <mvc:annotation-driven conversion-service="conversionService"></mvc:
annotation-driven>
</beans>
```

【程序解析】Spring MVC 框架提供了格式转换器，将类型转换工具配置到格式转换器中，再结合@RequestParam 注解实现了参数绑定过程中的类型转换。

12.5　Spring MVC 常见注解介绍

1. @RequestParam

（1）功能

用于将请求参数区数据映射到功能处理方法的参数上。例如：

Spring MVC 常见注解介绍

```
public String requestparam1(@RequestParam String username)
```

① 如果请求中包含 username 参数（如/requestparam1?username=zhang），则自动传入。

② 也可以使用@RequestParam("username")明确告诉 Spring Web MVC 使用 username 进行入参。

（2）主要参数

value：参数名字，即入参的请求参数名字，如 username 表示请求参数区中名为 username 的参数的值将传入。

required：是否必需，默认是 true，表示请求中一定要有相应的参数，否则将报 400 错误码。

defaultValue：表示如果请求中没有同名参数时的默认值，默认值可以是 SpEL 表达式，如 #{systemProperties["java.vm.version"]}。

例如：

```
public  String  test(@RequestParam(value="username",required=false)  String
username)
```

上述代码表示请求中可以没有名为 username 的参数，如果没有默认为 null，则需注意以下几点。

① 原子类型：必须有值，否则抛出异常，如果允许空值请使用包装类代替。

② Boolean 包装类型：默认 Boolean.FALSE，其他引用类型默认为 null。

例如：

```
public String requestparam5(@RequestParam(value="username", required=true, defaultValue="zhang") String username)
```

上述代码表示如果请求中没有名为 username 的参数，则默认值为“zhang”。如果请求中有多个同名的则又如何接收呢？给用户授权时，可能授予多个权限，先看如下代码：

```
public String test(@RequestParam(value="role") String roleList)
```

（3）如果请求参数类似 url?role=admin&role=user，则实际 roleList 入参的数据为“admin,user”，即多个数据之间使用“,”分割，应使用如下方式来接收多个请求参数：

```
public String test(@RequestParam(value="role") String[] roleList)
```

或

```
public String test(@RequestParam(value="list") List<String> list)
```

2. @PathVariable

功能：用于将请求 URL 中的模板变量映射到功能处理方法的参数上。

例如：

```
@RequestMapping(value="/users/{userId}/topics/{topicId}")
public String test( @PathVariable(value="userId") int userId, @PathVariable(value="topicId") int topicId)
```

如请求的 URL 为“控制器 URL/users/123/topics/456”，则自动将 URL 中的模板变量{userId}和{topicId}绑定到通过@PathVariable 注解的同名参数上，即入参后 userId=123、topicId=456。

3. @CookieValue

功能：用于将请求的 Cookie 数据映射到功能处理方法的参数上。

例如：

```
public String test(@CookieValue(value="JSESSIONID", defaultValue="") String sessionId)
```

以上配置自动将 JSESSIONID 值入参到 sessionId 参数上，defaultValue 表示 Cookie 中没有 JSESSIONID 时默认为空。

例如：

```
public String test2(@CookieValue(value="JSESSIONID", defaultValue="") Cookie sessionId)
```

传入的参数类型也可以是 javax.servlet.http.Cookie 类型。此外，@CookieValue 也拥有和@RequestParam 相同的 3 个参数，含义一样。

4. @RequestHeader

功能：用于将请求的头信息区数据映射到功能处理方法的参数上。

例如：

```
@RequestMapping(value="/header")
public String test(
    @RequestHeader("User-Agent") String userAgent,
```

```
        @RequestHeader(value="Accept") String[] accepts){
    }
```

以上配置自动将请求头“User-Agent”值入参到 userAgent 参数上，并将“Accept”请求头值入参到 accepts 参数上。此外，@RequestHeader 也拥有和@RequestParam 相同的 3 个参数，含义一样。

5. @ModelAttribute

功能如下。

① 绑定请求参数到命令对象。放在功能处理方法的入参上时，用于将多个请求参数绑定到一个命令对象，从而简化绑定流程，且自动暴露为模型数据，可在视图页面展示时使用。

② 暴露表单引用对象为模型数据：放在处理器的一般方法（非功能处理方法）上时，是为表单准备要展示的表单引用对象，如注册时需要选择的所在城市等，且在执行功能处理方法（@RequestMapping 注解的方法）之前，自动添加到模型对象中，在视图页面展示时使用。

③ 暴露@RequestMapping 方法返回值为模型数据。放在功能处理方法的返回值上时，是暴露功能处理方法的返回值为模型数据，在视图页面展示时使用。

6. @Value

功能：用于将一个 SpEL 表达式的结果映射到功能处理方法的参数上。

例如：

```
    public  String  test(@Value("#{systemProperties['java.vm.version']}")  String
jvmVersion)
```

7. @MatrixVariable

功能：用于接收 URL 的 path 中的矩阵参数。

语法格式：

```
    XXX/XXX/path;name=value;name=value
```

（1）如果是 XML 配置的 RequestMappingHandlerMapping，那么需要设置 removeSemicolon-Content 属性为 false。

（2）如果是注解的方式，直接设置<mvc:annotation-driven>的 enableMatrixVariables="true" 就可以了。

例如：测试的 URL 如下。

```
    http://localhost:9080/mvcexample/users/42;q=11;r=12/others/21;q=22;s=23
    @RequestMapping(value = "/users/{userId}/others/{otherUserId}",method =
    RequestMethod.GET)
    public void hello(
        //如果只有一个地方有 q，也可以这么取，但如果有多个 q，这样就错了，必须像第二个那样去
    指定取谁的 q 值
        // @MatrixVariable int q,
        @MatrixVariable(value="q", pathVar="userId") int q1,
        @MatrixVariable(value="q", pathVar="otherUserId") int q2,
        @MatrixVariable Map<String, String> matrixVars,
        @MatrixVariable(pathVar="userId") Map<String, String> userIdMatrixVars
```

```
    ) {
    // System.out.println("q=="+q);
    System.out.println("q1="+q1+",q2="+q2+",matrixVars="+matrixVars+",userIdMatr
    ixVars="+userIdMatrixVars);
    }
```

运行结果如下。

```
    q1=11,q2=22,matrixVars={q=[11, 22], r=[12], s=[23]},userIdMatrixVars={q=[11],
r=[12]}
```

12.6 REST 简介

REST 简介

表征状态转移（Representational State Transfer，REST）是罗伊・菲尔丁（Roy Fielding）于 2000 年在其博士论文中提出来的一种软件架构风格。

REST 从资源的角度观察整个网络，分布在各处的资源由 URI 确定，而客户端的应用通过 URI 来获取资源的表征。获得这些表征会使这些应用程序转变其状态。随着不断获取资源的表征，客户端应用不断地转变其状态，这就是所谓的表征状态转移。

如果一个架构符合 REST 原则，我们就称它为 RESTful 架构。

1. REST 的特点

（1）REST 是设计风格而不是标准。REST 通常基于使用 HTTP、URI、XML 以及 HTML 这些现有的广泛流行的协议和标准。

（2）资源由 URI 来指定。

（3）对资源的操作包括获取、创建、修改和删除资源。

（4）通过操作资源的表现形式来操作资源。

2. REST 的优点

（1）可以利用缓存 Cache 来提高响应速度。

（2）通信本身的无状态性可以让不同的服务器处理一系列请求中的不同请求，提高服务器的扩展性。

（3）浏览器即可作为客户端，简化软件需求。

（4）相对于其他叠加在 HTTP 协议之上的机制，REST 的软件依赖性更小。

（5）不需要额外的资源发现机制。

（6）在软件技术演进中的兼容性更好。

本章小结

本章主要对 Spring MVC 的核心注解的应用进行详细讲解，介绍了 Controller 和 RequestMapping 注解类型的相关知识。通过对本章的学习，读者能够了解 Spring MVC 核心注解的作用，并掌握 Spring MVC 常用注解的使用方法。

习题

1. 下列属性（　　）可用于@RequestMapping 注解匹配参数值。

   A. value　　B. name　　C. params　　D. header

2. 简述@Controller 注解的使用方式。
3. 列举@RequestParam 注解的使用方式。

上机指导

创建一个 Web 工程，并引入 Spring 框架，完成模拟用户登录功能。

13 第13章 数据验证

学习目标

- 掌握声明式数据验证的基本用法
- 掌握验证中常见的注解方法
- 掌握验证中的错误消息处理方法
- 掌握 Spring MVC 中的统一异常处理方法

13.1 声明式数据验证的基本用法

在项目开发的过程中，数据的验证几乎无处不在，在前后台数据之间，控制层、表现层和模型层之间，以及开发团队之间的数据调用等多个环节，都需要对数据进行格式或数值的检查验证，以确保程序正常运行。

声明式数据验证的基本用法

从 Spring 3 版本开始，Spring MVC 框架提供了 JSR-303 验证框架。JSR-303 是 Java EE 6 中的一项子规范，叫作 Bean Validation。随着技术的进步，出现了一个专门进行数据验证的框架 Hibernate Validator。该框架在 JSR-303 基础上提供了更全面的验证支持注解，目前应用较为广泛。使用 Hibernate Validator 验证框架需要提供特定的支持包环境。接下来我们通过登录验证案例，来了解数据验证的具体方式。

【实例 13.1】使用 Spring MVC 框架结合 Hibernate Validator 进行用户登录验证。

这里暂时不使用数据库操作。具体开发方式如下（代码详见 Springmvc_Ch13\WEB-INF\web.xml；Springmvc_Ch13\WEB-INF\springmvc-servlet.xml；Springmvc_Ch13\login.jsp；Springmvc_Ch13\success.jsp；Springmvc_Ch13\src\com\inspur\action\LoginController.java；Springmvc_Ch13\src\com\inspur\entity\Users.java）。

（1）新建 Web 项目及登录验证的目录结构，如图 13-1 所示。

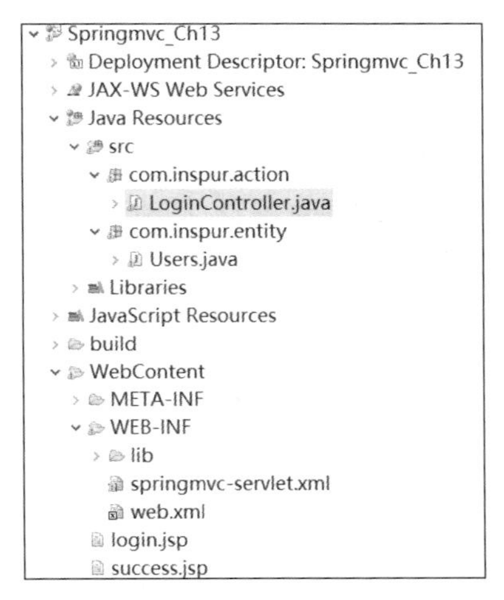

图 13-1 登录验证代码组织结构

（2）添加 Hibernate-validator 的 JAR 包，文件为 hibernate-validator-4.3.0.Final-dist.zip，该支持包可从 Hibernate 官网下载。

- validation-api-1.0.0.GA.jar（JSR-303 规范的 API 包）。
- hibernate-validator-4.3.0.Final.jar（Hibernate 对 JSR-303 的扩展支持包）。
- jboss-logging-3.1.0.CR2.jar（Formatter SPI 简介）。

（3）在 Spring MVC 的配置文件中配置注解扫描和加载注解驱动。

```
<?xml version="1.0" encoding="UTF-8"?>
<beans xmlns="http://www.springframework.org/schema/beans"
       xmlns:xsi="http://www.w3.org/2001/XMLSchema-instance"
       xmlns:context="http://www.springframework.org/schema/context"
       xmlns:mvc="http://www.springframework.org/schema/mvc"
       xsi:schemaLocation="http://www.springframework.org/schema/beans
       http://www.springframework.org/schema/beans/spring-beans.xsd
       http://www.springframework.org/schema/context
       http://www.springframework.org/schema/context/spring-context.xsd
       http://www.springframework.org/schema/mvc
       http://www.springframework.org/schema/mvc/spring-mvc.xsd">
       <context:component-scan base-package="com.inspur.action">
       </context:component-scan>
    <mvc:annotation-driven></mvc:annotation-driven>
</beans>
```

完成 Hibernate validator 的数据验证功能，需要通过 Spring MVC 框架提供的 LocalValidatorFactoryBean 类生成验证器，来实现 Hibernate Validator 框架中的相关接口。在 Spring 5 版本中，通过<mvc:annotation-driven>标签引入注解驱动，其中内置了该实现类，从而简化了该类的实例配置。如果需要更改该实现类中的一些默认属性信息，可通过重新配置其中的属性值来完成。

（4）完成实体类 Users 的开发，代码如下。

```
package com.inspur.entity;
import org.hibernate.validator.constraints.NotBlank;
```

```
public class Users {
    @NotBlank
    private String name;
    private String password;
    public String getName() {
        return name;
    }
    public void setName(String name) {
        this.name = name;
    }
    public String getPassword() {
        return password;
    }
    public void setPassword(String password) {
        this.password = password;
    }
}
```

Spring MVC 框架通过参数绑定机制，可以将前端 JSP 页面请求中的数据，绑定到实体对象中。在实体对象中，通过@NotBlank 注解，实现对实体对象属性值的验证，Users 对象的 name 属性值在绑定时如果为空格，则会触发报错机制。

（5）完成对控制器 LoginControlleer 的开发，代码如下。

```
@Controller
public class LoginController{
    @RequestMapping("/loginCheck.action")
    public ModelAndView loginCheck( @Valid Users u, BindingResult br
            ) throws Exception {
        ModelAndView mav=new ModelAndView();
        List<ObjectError> errors=br.getAllErrors();
        for(int i=0;i<errors.size();i++){
            System.out.println(errors.get(i).getDefaultMessage());
        }
        if("123".equals(u.getPassword())){
            mav.setViewName("success.jsp");
        }
        return mav;
    }
}
```

Spring MVC 框架通过参数绑定机制将 JSP 页面请求中的数据，直接封装到 Users 对象 u 中。通过@Valid 注解实现对该对象的验证工作，结合实体类中的具体验证内容，共同完成对 u 对象中属性值的验证。如果验证过程不符合验证规则，错误信息将会被封装到 BindingResult 对象 br 中，通过方法可以获取其中的错误信息列表。这些错误信息都是在导入的 JAR 包中定义好的格式，如果需要自定义格式，可以修改注解的某些属性（这里按默认的错误信息显示）。特别需要注意的是，BindingResult 对象的声明必须紧邻@Valid 注解的对象之后。

（6）完成登录页面 login.jsp 的开发，代码如下。

```
<body>
   <h3>登录页面</h3>
   <form action="/Springmvc_Ch13/loginCheck.action" method="post">
       请输入用户名：<input type="text" name="name"/><br/>
```

```
        请输入密码：<input type="password" name="password"/><br/>
        <input type="submit" value="点击登录"/>
    </form>
</body>
```

【程序解析】在输入用户名和密码时，input 元素的 name 属性值，必须与实体类 Users 中的属性名一致，才能完成数据与控制器中 Users 对象的绑定。

（7）代码测试。用户名中没有输入数据就提交时，控制层中得到验证的匹配结果如图 13-2 和图 13-3 所示。

登录页面

请输入用户名：

请输入密码：···

点击登录

图 13-2　登录页面测试

二月 20, 2020 11:25:54 下午 org

信息: Looking for @Controller

二月 20, 2020 11:25:54 下午 org

信息: FrameworkServlet 'sprin

不能为空

图 13-3　控制器后台匹配结果

13.2　数据验证中常见的注解

数据验证中常见的注解

Hibernate Validator 提供了大量的数据验证注解，这里只列举常见的一部分，详细内容可以参考 Hibernate Validator 官方文档了解其他验证注解及进行自定义的验证注解。

（1）@AssertFalse

验证注解的元素值是否为 false。

验证的数据类型：Boolean、boolean。

（2）@AssertTrue

验证注解的元素值是否为 true。

验证的数据类型：Boolean、boolean。

（3）@NotNull

验证注解的元素值不是 null。

验证的数据类型：任意类型。

（4）@Null

验证注解的元素值是 null。

验证的数据类型：任意类型。

（5）@Min(value=值)

验证注解的元素值大于等于@Min 指定的 value 值。

验证的数据类型：BigDecimal、BigInteger、byte、short、int、long 等任意 Number 或 CharSequence（存储的是数字）子类型。

（6）@Max(value=值)

验证注解的元素值小于等于@Max 指定的 value 值。

验证的数据类型：和@Min 要求一样。

（7）@DecimalMin(value=值)

验证注解的元素值大于等于@DecimalMin 指定的 value 值。

验证的数据类型：和@Min 要求一样。

（8）@DecimalMax(value=值)

验证注解的元素值小于等于@DecimalMax 指定的 value 值.

验证的数据类型：和@Min 要求一样。

（9）@Digits(integer=整数位数,fraction=小数位数)

验证注解的元素值的整数位数和小数位数上限。

验证的数据类型：和@Min 要求一样。

（10）@Size(min=下限,max=上限)

验证注解的元素值在 min 和 max（包含）指定区间之内，如字符长度、集合大小。

验证的数据类型：字符串、Collection、Map、数组等。

（11）@Past

验证注解的元素值（日期类型）比当前时间早。

验证的数据类型：java.util.Date、java.util.Calendar、Joda Time 类库的日期类型。

（12）@Future

验证注解的元素值（日期类型）比当前时间晚。

验证的数据类型：与@Past 要求一样。

（13）@NotEmpty

验证注解的元素值不为 null 且不为空字符串（字符串长度不为 0、集合大小不为 0）。

验证的数据类型：CharSequence 子类型、Collection、Map、数组。

（14）@NotBlank

验证注解的元素值不为空（不为 null、去除首位空格后长度为 0），不同于@NotEmpty，@NotBlank 只应用于字符串且在比较时会去除字符串的首位空格。

验证的数据类型：CharSequence 子类型。

（15）@Length(min=下限,max=上限)

验证注解的元素值长度在 min 和 max 的区间内。

验证的数据类型：CharSequence 子类型。

（16）@Range(min=最小值, max=最大值)

验证注解的元素值在最小值和最大值之间。

验证的数据类型：BigDecimal、BigInteger、CharSequence、byte、short、int、long 等原子类型和包装类型。

（17）@Email(regexp=正则表达式,flag=标志的模式)

验证注解的元素值是 E-mail 邮箱，也可以通过 regexp 和 flag 指定自定义的 email 格式。

验证的数据类型：CharSequence 子类型（如 String）。

（18）@Pattern(regexp=正则表达式,flag=标志的模式)

验证注解的元素值与指定的正则表达式匹配。

验证的数据类型：String、任何 CharSequence 的子类型。

（19）@Valid

指定递归验证关联的对象：如用户对象中有一个地址对象属性，如果想在验证用户对象时一起验证地址对象，可在地址对象上加@Valid 注解进行级联验证。

验证的数据类型：任何非原子类型。

13.3　数据验证中错误消息的处理

数据验证中错误消息的处理

当验证出错时，我们需要给用户展示错误消息并告诉用户出错的原因，因此要为验证约束注解指定错误消息。错误消息可通过验证约束注解的 message 属性指定。验证约束注解指定错误消息有如下两种方式。

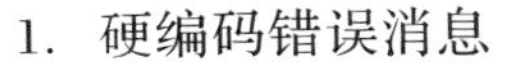

1. 硬编码错误消息

直接在验证约束注解上指定错误消息，如下所示。

```
@NotNull(message = "用户名不能为空")
@Length(min=5, max=20, message="用户名长度必须在 5-20 之间")
@Pattern(regexp = "^[a-zA-Z_]\\w{4,19}$", message = "用户名必须以字母和下画线开头，
可由字母、数字、下画线组成")
private String username;
```

错误消息使用硬编码指定，在有些场景是不适用的。如在国际化场景下，需要对不同的国家显示不同的错误消息。另外在更换错误消息时是比较麻烦的，需要找到相应的类进行更换，并重新编译发布。

2. 从资源消息文件中根据消息键读取错误消息

直接使用 Spring MVC 框架提供的 MessageSource Bean 进行消息的匹配和管理，在配置文件中将修改后的消息实例重新加载到验证器中。

13.4　Spring MVC 的统一异常处理

Spring MVC 的统一异常处理

在使用 Spring MVC 进行项目开发时，不管是对底层的数据库操作过程，还是业务层的处理过程，或者是控制层的处理过程，都会不可避免地遇到各种可预知或不可预知的异常。如果每个异常都要进行单独处理，则会造成系统的代码耦合度高，工作量大且不统一，维护的工作量也很大。

为了将各种类型的异常处理从各处理过程解耦出来，实现异常信息的统一处理和维护，Spring MVC 提供了如下 3 种处理异常的方式。

① 使用 Spring MVC 提供的简单异常处理 SimpleMappingExceptionResolver。

② 实现 Spring MVC 的异常处理接口 HandlerExceptionResolver，自定义异常处理器。

③ 使用@ExceptionHandler 注解实现异常处理。

这里使用实现 HandlerExceptionResolver 接口的方式自定义异常处理。

【实例 13.2】 通过退出功能实现统一异常处理。

这里暂时不使用数据库操作。具体开发方式如下（代码详见 Springmvc_Ch13\WEB-INF\web.xml；Springmvc_Ch13\WEB-INF\springmvc-servlet.xml；Springmvc_Ch13\login.jsp；Springmvc_Ch13\error.jsp；Springmvc_Ch13\src\com\inspur\action\LoginController.java；Springmvc_Ch13\src\com\inspur\exception\MyException.java）。

（1）在源代码中新添加 com.inspur.exception.MyException 类，代码如下。

```
public class MyExcepition implements HandlerExceptionResolver{
    public ModelAndView resolveException(HttpServletRequest request,
        HttpServletResponse response, Object obj, Exception ex) {
        System.out.println("进入了我的异常处理器");
        ModelAndView mav=new ModelAndView();
        mav.addObject("errorInfo", ex.getMessage());
        mav.setViewName("error.jsp");
        return mav;
    }
}
```

该类必须实现 Spring MVC 框架提供的 HandlerExceptionResolver 接口，在其实现的方法中，异常的具体信息被封装到 Exception 对象 ex 中，可以直接用来获取错误信息和其他信息。

（2）在 Spring MVC 的配置文件中配置自定义异常类的实例，代码如下。

```
<?xml version="1.0" encoding="UTF-8"?>
<beans xmlns="http://www.springframework.org/schema/beans"
        xmlns:xsi="http://www.w3.org/2001/XMLSchema-instance"
        xmlns:context="http://www.springframework.org/schema/context"
        xmlns:mvc="http://www.springframework.org/schema/mvc"
        xsi:schemaLocation="http://www.springframework.org/schema/beans
        http://www.springframework.org/schema/beans/spring-beans.xsd
        http://www.springframework.org/schema/context
        http://www.springframework.org/schema/context/spring-context.xsd
        http://www.springframework.org/schema/mvc
        http://www.springframework.org/schema/mvc/spring-mvc.xsd">
    <context:component-scan base-
    package="com.inspur.action"></context:component-scan>
    <mvc:annotation-driven></mvc:annotation-driven>
    <bean class="com.inspur.exception.MyExcepition"></bean>
</beans>
```

（3）在控制器 LoginController 类中添加一个退出登录的方法，代码如下。

```
@RequestMapping("/logout.action")
public ModelAndView logout(HttpSession hs) throws Exception{
    hs.removeAttribute("");
    System.out.println(9/0);
    ModelAndView mav=new ModelAndView();
    mav.setViewName("redirect:login.jsp");
    return mav;
    }
```

【程序解析】 这里使用一个算术运行时的异常来模拟方法执行过程中，可能遇到的各种类型异常，当异常发生时，该方法会将异常向上抛出，交给统一异常处理器 MyExcepition 类进行处理。

（4）在 login.jsp 页面中增加一个退出的表单，代码如下。

```
<form action="/Springmvc_Ch13/logout.action" method="post">
        <input type="submit" value="安全退出"/>
</form>
```

（5）增加一个错误显示页面 error.jsp，代码如下。

```
<body>
      <h2>程序运行有误，请重新登录</h2>
      <h3>错误内容：${errorInfo}</h3>
</body>
```

当请求从统一异常处理类 MyExcepition 进入错误页面时，可使用 EL 表达式获取封装的错误信息。

（6）测试运行结果如图 13-4 所示。

程序运行有误，请重新登录

错误内容：/ by zero

图 13-4　统一异常检测结果

本章小结

本章主要介绍了声明式数据验证的用法以及验证中常见的注解方法、验证中错误消息的处理方法、统一异常处理的方法。

习题

1. 完成声明式数据验证的操作步骤有哪些？主要用到了哪些注解？
2. 完成 Spring MVC 统一异常处理的步骤有哪些？

上机指导

创建一个 Web 工程，并引入 Spring 框架和 Hibernate Validator 完成用户信息的验证功能（用户名不能为空，年龄为整数）。

14 第14章　拦截器

学习目标

- 了解拦截器的概念及其常见应用
- 掌握拦截器接口的实现方法
- 掌握拦截器的配置方法
- 掌握拦截器的执行流程（正常流程）
- 理解常见应用的性能监控

14.1　拦截器概述

14.1.1　拦截器的应用

拦截器概述

Spring MVC 的处理器拦截器类似 Servlet 开发中的过滤器 Filter，用于对处理器进行预处理和后处理。拦截器的本质是 AOP（面向切面编程），也就是说符合横切关注点的所有功能都可以放入拦截器中实现。拦截器在企业级的项目中，发挥着巨大的作用。下面是几个拦截器常见的业务场景。

（1）日志记录：记录请求信息的日志，以便进行信息监控、信息统计、计算及页面访问量等。

（2）权限检查：如登录检测，进入处理器检测是否登录，如果没有登录，就直接返回到登录页面。

（3）性能监控：有时候系统在某段时间运行速度突然变慢，这时可以使用拦截器在进入处理器之前记录开始时间，在处理完成后记录结束时间，从而得到该请求的处理时间（如果有反向代理，如 apache，则可以自动记录）。

（4）通用行为：读取 Cookie 得到用户信息并将用户对象放入请求，从而方便后续流程使用，还有如提取国际化、主题信息等，只要是多个处理器都需要的就可使用拦截器实现。

14.1.2 Spring MVC 架构中的拦截器

在图 14-1 中，拦截器是在请求映射过程中被调用的。Spring MVC 收到用户请求后，前端控制器会调用处理器映射器进行请求的映射，处理器映射器处理的结果是一个 HandlerExecutionChain（处理器执行链）对象，该对象中包含一个 Handler 处理器（页面控制器）对象、多个 HandlerInterceptor（拦截器）。

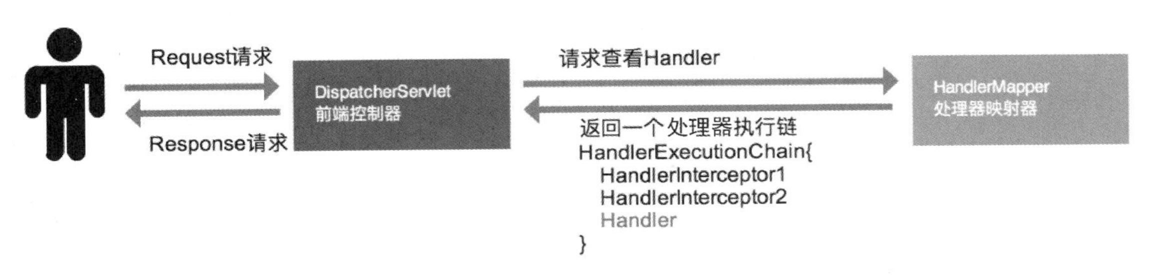

图 14-1 Spring MVC 架构中的拦截器

14.2 拦截器的实现

14.2.1 拦截器的定义

Spring MVC 中提供了一个 HandlerInterceptor 接口实现拦截器功能，接口内容如下。

```
public interface HandlerInterceptor {
    boolean preHandle(HttpServletRequest request,
         HttpServletResponse  response, Object handler) throws Exception;
    void postHandle(HttpServletRequest request,
         HttpServletResponse response, Object handler,
          ModelAndView modelAndView) throws Exception;
    void afterCompletion(HttpServletRequest request,
         HttpServletResponse response, Object handler,
         Exception ex) throws Exception;
}
```

该接口提供了 3 个未实现的方法。

（1）preHandle：预处理回调方法，实现处理器的预处理。

该方法是最先执行的方法，在处理器方法执行之前执行，通常用于登录验证、数据验证等。其返回值为布尔类型，true 表示继续流程（如调用下一个拦截器或处理器），false 表示流程中断（如登录检查失败），不会继续调用其他的拦截器或处理器，此时我们需要通过第 2 个参数 response 来产生响应。

（2）postHandle：后处理回调方法，实现处理器的后处理。

该方法在处理器方法执行完成后、视图解析之前执行，此方法通常用于对模型数据和视图进行进一步的修改。我们可以通过 modelAndView（模型和视图对象）对模型数据或视图进行处理，modelAndView 也可能为 null。

（3）afterCompletion：整个请求处理完毕回调方法，在视图渲染完毕时回调。

由于该方法是在整个请求处理完毕后才执行的，故而可用来进行资源的清理、日志记录等。该方法类似 try-catch-finally 中的 finally，不过处理器执行链中只有 preHandle 方法返回值为 true 的拦截器的 afterCompletion 方法会被调用。

如果要自定义一个拦截器，只要让它继承 HandlerInterceptor 接口，实现以上 3 个未实现的方法即可。如下所示。

```
public class MyInterceptor implements HandlerInterceptor{
    public boolean preHandle(HttpServletRequest request,
            HttpServletResponse response, Object handler) throws Exception {
            //预处理回调方法，实现处理器的预处理，可以用于完成登录验证，数据验证等
            return false;
    }
    public void postHandle(HttpServletRequest request,
            HttpServletResponse response, Object handler,
            ModelAndView mv) throws Exception {
            //后处理回调方法，实现处理器的后处理，可以对模型数据和视图进行进一步的修改
    }
    public void afterCompletion(HttpServletRequest request,
           HttpServletResponse response, Object handler,
           Exception ex)throws Exception {
           //请求处理完毕回调方法，通常用来完成资源的关闭或记录日志等操作
    }
}
```

但是在某些情况下，我们可能只需要实现 3 个回调方法中的某一个，例如登录验证，我们只要在用户发出指向网站内容的请求时，验证用户是否已经登录，后处理回调方法和请求处理完毕回调方法不需要进行任何处理。如果通过继承 HandlerInterceptor 接口的方式实现的话，不管是否需要，3 个方法都必须实现。针对这种情况，Spring 提供了一个 HandlerInterceptor- Adapter 适配器（一种适配器设计模式的实现）。我们想要只实现必需的回调方法，只需继承这个抽象类即可。如下面的案例所示。

```
public class MyInterceptor extends HandlerInterceptorAdapter{
    public boolean preHandle(HttpServletRequest request,
        HttpServletResponse response,Object handler) throws Exception {
        System.out.println("MyInterceptor 的 preHandle 方法正在被执行……");
        return true;
    }
}
```

14.2.2 拦截器的配置

拦截器需要在前端控制器中进行配置才能生效。<mvc:interceptors/>标签用于配置一组拦截器，标签内部可以使用多个<mvc:interceptor/>或直接使用<bean/>标签分别配置每个拦截器的信息。

拦截器配置

在<mvc:interceptors/>标签内部，直接使用<bean/>标签配置的拦截器对所有的请求都生效。

如果需要指定拦截器生效的路径，需要使用<mvc:interceptor/>标签标识一个拦截器的配置信息。<mvc:interceptor/>的子标签<mvc:mapping/>用于配置拦截器生效的路径，其 path 属性用于指定请求路径，“/**”表示拦截所有路径。如果需要在指定路径中排除一些路径，可以使用<mvc:exclude-mapping>子标签。不过，使用<mvc:exclude-mapping>时必须保证<mvc:mapping>标签存在。子标签<bean>可用来配置拦截器实现类的位置。

在下面的代码示例中，配置了两个拦截器：第一个拦截器（FirstInterceptor）的<bean/>标签未经包装直接配置在<mvc:interceptors/>内部，对所有请求有效；而第二个拦截器（SecondInterceptor）经过< mvc:interceptor/>标签的包装设置对除/login 请求之外的所有请求有效。

```
<!-- 配置拦截器 -->
<mvc:interceptors>
   <!-- 直接配置在 mvc:interceptors 内部的拦截器对所有请求有效 -->
    <bean class="com.inspur.interceptor.FirstInterceptor"></bean>
    <mvc:interceptor>
         <!-- 配置拦截器生效的路径 -->
         <mvc:mapping path="/**"/>
         <!-- 配置拦截器不生效的路径，在生效的路径范围内，消除不生效的路径 -->
         <mvc:exclude-mapping path="/login"/>
         <!-- 拦截器的位置 -->
         <bean class="com.inspur.interceptor.SecondInterceptor"></bean>
    </mvc:interceptor>
</mvc:interceptors>
```

需要注意的是，<mvc:interceptor>的 3 个子标签是有严格的顺序要求的，必须按照<mvc:mapping>→<mvc:exclude-mapping>→<bean>的顺序进行配置，否则程序会报错。

14.3　拦截器的使用

在介绍拦截器的案例之前，我们先准备一个用户登录模块：用户在登录页面录入用户名、密码，单击“登录”按钮将请求提交给处理器，处理器对用户信息进行验证，验证通过后进入网站主界面，验证失败则会返回登录页面，并提示消息“用户名或密码错误，请重新登录!”。相关的代码准备可扫描二维码查看。

代码准备

【实例 14.1】模拟完成用户登录模块。

（1）配置 WEB-INF/web.xml

```
<!-- Spring MVC 前端控制器 DispatcherServlet -->
<servlet>
  <servlet-name>springmvc</servlet-name>
  <servlet-class>org.springframework.web.servlet.DispatcherServlet</servlet-class>
  <load-on-startup>1</load-on-startup>
</servlet>
<servlet-mapping>
     <servlet-name>springmvc</servlet-name>
     <url-pattern>/</url-pattern>
</servlet-mapping>
```

```
<!--字符编码过滤器 -->
<filter>
    <filter-name>CharacterEncodingFilter</filter-name>
    <filter-class>org.springframework.web.filter.CharacterEncodingFilter
    </filter-class>
    <init-param>
        <param-name>encoding</param-name>
        <param-value>UTF-8</param-value>
    </init-param>
</filter>
<filter-mapping>
    <filter-name>CharacterEncodingFilter</filter-name>
    <url-pattern>/*</url-pattern>
</filter-mapping>
```

（2）开发前端控制器 WEB-INF/springmvc-servlet.xml

```
<?xml version="1.0" encoding="UTF-8"?>
<beans xmlns="http://www.springframework.org/schema/beans"
    xmlns:xsi="http://www.w3.org/2001/XMLSchema-instance"
    xmlns:context="http://www.springframework.org/schema/context"
    xmlns:aop="http://www.springframework.org/schema/aop"
    xmlns:mvc="http://www.springframework.org/schema/mvc"
    xsi:schemaLocation="http://www.springframework.org/schema/beans
        http://www.springframework.org/schema/beans/spring-beans.xsd
        http://www.springframework.org/schema/context
        http://www.springframework.org/schema/context/spring-context.xsd
        http://www.springframework.org/schema/aop
        http://www.springframework.org/schema/aop/spring-aop.xsd
        http://www.springframework.org/schema/mvc
        http://www.springframework.org/schema/mvc/spring-mvc.xsd">
    <!-- 指定自动扫描的包，客户端发出请求时，mvc 自动在该包内查找 url 请求对应的处理方法 -->
    <context:component-scan baspackage="com.inspur.controller" />
    <!-- 视图解析器 -->
    <bean class="org.springframework.web.servlet.view.InternalResourceViewResolver">
        <property name="prefix" value="/page/"></property>
        <property name="suffix" value=".jsp"></property>
    </bean>
</beans>
```

（3）开发用户类 com.inspur.po.User

设计该网站用户包含 id（用户编码）、name（姓名）、password（密码）、access（是否持有网站访问许可）4 项基本信息。下面的案例代码中省略了各属性的 getXXX 和 setXXX 方法，读者练习时，要注意添加。

```
public class User {
    private String id;
    private String name;
    private String password;
    private String access;
    …
    public User(){
    }
    public User(String id, String name, String password,String access){
        this.id = id;
        this.name = name;
        this.password = password;
```

```
            this.access = access;
        }
    }
```

为了方便在处理器中添加新用户，该类中添加了含参数的构造函数，此时 Java 不会再默认产生无参构造函数，必须进行手动添加。否则，Spring MVC 进行数据绑定时会报错。

（4）开发网站登录页面 page/login.jsp

在登录页面上，使用 EL 表达式获取了一个 msg 对象的值，留作将来登录不成功的提示消息展示之用。

```
<%@ page language="java" import="java.util.*" pageEncoding="UTF-8"%>
<!DOCTYPE HTML PUBLIC "-//W3C//DTD HTML 4.01 Transitional//EN">
<html>
<head>
  <title>用户登录</title>
</head>
<body>
${msg}<br>
<form action="dologin" method="post">
     姓名：<input type="text" name="username" /><br>
     密码：<input type="text" name="password" /><br>
     <input type="submit" value="登录">
</form>
</body>
</html>
```

（5）开发网站主页面 page/main.jsp

```
<%@ page language="java" import="java.util.*" pageEncoding="UTF-8"%>
<!DOCTYPE HTML PUBLIC "-//W3C//DTD HTML 4.01 Transitional//EN">
<html>
  <head>
     <title>浪潮集团欢迎您</title>
  </head>
  <body>
  <% System.out.println("网站主页面加载成功……"); %>
     <h3>浪潮集团</h3>
     <p align="left">欢迎您：${username}</p>
  </body>
</html>
```

（6）开发处理器 com.inspur.controller.LoginController.java

在该文件中，我们使用静态的 List 集合模拟数据库中存储的用户信息。dologin 方法接收到用户的用户名和密码后会去数据库中查询是否为合法用户，如果是，就调用进入主页面的请求进入网站；否则，将请求转发到登录页面，要求用户重新输入合法的用户名和密码。

```
@Controller
public class LoginController {
```

```
    static List<User> all_user = new ArrayList<User>(Arrays.asList(new User("1",
        "张三", "123","1"),new User("2", "李四", "123456", "1"),new User("3",
        "王五", "123456", "0")));

    @RequestMapping(value="/login")
    public String gologin(){
        return "login";
    }
    @RequestMapping(value="/dologin")
    public String dologin(String username, String password, HttpSession session,
        Model model){
        if(username != null && !"".equals(username)&&
           password != null && !"".equals(password)){
             for(User u:all_user){
                  if(u.getName().equals(username)&&u.getPassword().equals
                  (password)){
                       session.setAttribute("username", username);
                       session.setAttribute("userid", u.getId());
                       session.setAttribute("useraccess", u.getAccess());

                       return "forward:gomain";
                  }
             }
        }
        model.addAttribute("msg", "用户名或密码错误，请重新登录！");
        return "forward:login";
    }

    @RequestMapping(value="/gomain")
    public String gomain(){
         System.out.println("控制器方法正在执行……");
         return "main";
    }
}
```

14.3.1 单个拦截器的执行流程

单个拦截器的执行流程

在前面对 HandlerInterceptor 接口的介绍中提到，该接口的第三个方法的执行是有一定顺序的，接下来我们对拦截器中每个方法的执行时机进行介绍。

为防止用户绕过登录过程，通过直接在浏览器中输入主界面地址的方式进入网站，可以为上述案例添加用户登录状态检测拦截器。拦截器验证用户在当前会话中是否已经登录过，如果是，可以继续浏览网站；否则，网页跳转到登录页面，并提示消息“您尚未登录，不允许直接访问网站!”。

【实例 14.2】模拟添加检测用户登录状态的拦截器。

1. 定义拦截器 com.inspur.interceptor.LoginInterceptor

在这个用户登录状态拦截检查的案例中，实际上只用到了请求预处理方法。下面给读者演示各方法的执行时机，我们在各方法中添加了向控制台输出的信息，以便后面进行查看及对比。

```
public class LoginInterceptor implements HandlerInterceptor{
    /**
     * 最先执行的方法，在处理器方法执行之前执行
     * 作用：进行登录验证、数据验证等
     * 返回值是布尔类型：true 表示继续进行下一步（执行下一个拦截器或者处理器方法）
     */
    public boolean preHandle(HttpServletRequest request, HttpServletResponse
        response,Object handler) throws Exception {
        HttpSession session = request.getSession();
        if(session.getAttribute("userid")!=null){
            System.out.println("1.LoginInterceptor 的 preHandle 方法正在被执
            行……[请求通过拦截器检查]");
            return true;
        }else{
            System.out.println("1.LoginInterceptor 的 preHandle 方法正在被执
            行……[请求被拦截]");
            request.setAttribute("msg", "您尚未登录，不允许直接访问网站！");
            request.getRequestDispatcher("/login").forward(request,
            response);
        }
        return false;
    }
    /**
    * 执行时机：处理器方法执行完之后、视图解析之前执行
    * 作用：对请求中的模型数据和视图进行进一步的修改
    */
    public void postHandle(HttpServletRequest request, HttpServletResponse
        response,Object handler, ModelAndView mv) throws Exception {
        System.out.println("1.LoginInterceptor 的 postHandle 方法正在被执行……");
    }
    /**
     * 在整个请求处理完毕后执行
     * 通常用来进行资源的关闭或记录日志等操作
     */
    public void afterCompletion(HttpServletRequest request,
        HttpServletResponse response, Object hanler, Exception exp)
        throws Exception {
        System.out.println("1.LoginInterceptor 的 afterCompletion 方法正在被执
        行……");
    }
}
```

2. 在前端控制器 WEB-INF/springmvc-servlet.xml 中添加拦截器配置信息

在这个案例中，我们配置用户登录状态拦截器生效的路径是除 login 相关请求之外的所有请求。

```
<!-- 配置拦截器 -->
<mvc:interceptors>
    <mvc:interceptor>
    <!-- 配置拦截器生效的路径 -->
    <mvc:mapping path="/**"/>
```

```
        <!-- 配置拦截器不生效的路径，在生效的路径范围内，消除不生效的路径 -->
        <mvc:exclude-mapping path="/*login"/>
        <!-- 拦截器的位置 -->
        <bean class="com.inspur.interceptor.LoginInterceptor"></bean>
        </mvc:interceptor>
    </mvc:interceptors>
```

3. 执行测试

（1）绕过登录过程，直接访问网站主界面进行测试

在这个案例中，用户登录状态拦截器生效的路径是除 login 相关请求之外的所有请求。首先尝试直接访问网站主界面。

请求地址：http://127.0.0.1:8080/mvc-interceptor/gomain。

页面显示如图 14-2 所示。

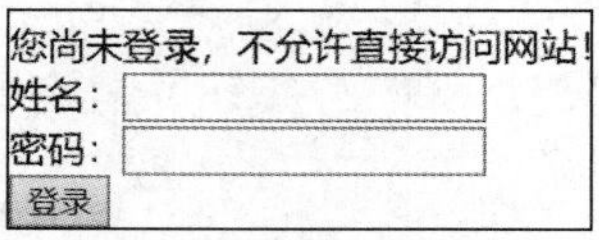

您尚未登录，不允许直接访问网站！
姓名：
密码：
登录

图 14-2 未登录用户访问网站的页面显示

控制台显示如图 14-3 所示。

```
1.LoginInterceptor的preHandle方法正在被执行...[请求被拦截]
```

图 14-3 未登录用户访问网站的控制台显示

（2）按照正常登录流程进行登录，输入用户名（张三）及密码（123）。

请求地址：http://127.0.0.1:8080/mvc-interceptor/login。

页面显示如图 14-4 所示。

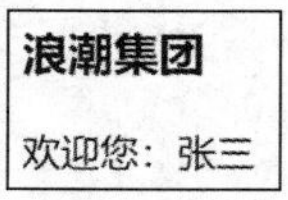

浪潮集团

欢迎您：张三

图 14-4 按照正常登录流程访问网站的页面显示

控制台显示如图 14-5 所示。

```
1.LoginInterceptor的preHandle方法正在被执行...[请求通过拦截器检查]
控制器方法正在执行...
1.LoginInterceptor的postHandle方法正在被执行...
网站主页面加载成功...
1.LoginInterceptor的afterCompletion方法正在被执行...
```

图 14-5 按照正常登录流程访问的控制台显示

（3）当前会话中登录过之后，直接访问网站主页面。

请求地址：http://127.0.0.1:8080/mvc-interceptor/gomain。

页面显示如图 14-6 所示。

浪潮集团

欢迎您：张三

图 14-6 保存会话后访问网站的页面显示

控制台显示如图 14-7 所示。

```
1.LoginInterceptor的preHandle方法正在被执行...[请求通过拦截器检查]
控制器方法正在执行...
1.LoginInterceptor的postHandle方法正在被执行...
网站主页面加载成功...
1.LoginInterceptor的afterCompletion方法正在被执行...
```

图 14-7 保存会话后访问网站的控制台显示

至此，测试完成。拦截器根据用户的登录状态成功地对试图进入的请求进行了放行或拦截。根据测试结果，不难看出单个拦截器中方法的执行流程如图 14-8 所示。

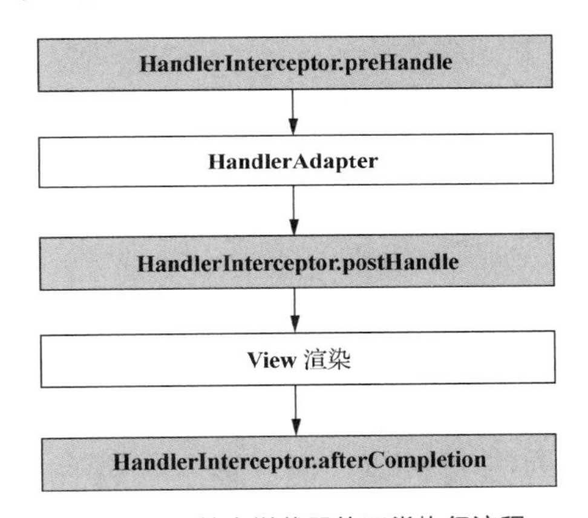

图 14-8 单个拦截器的正常执行流程

用户请求被拦截，预回调方法返回 false 时，会根据预处理方法中的设定进行请求的处理，图 14-8 中 preHandle 之后的所有后续步骤在本次请求中不再继续进行。

14.3.2 多个拦截器的执行流程

多个拦截器的执行流程

在大型的企业级项目中，通常不会只存在一个拦截器，而是通过多个拦截器相互配合进行不同功能的拦截过滤。多个拦截器组成一个拦截器链，接下来，我们通过一个案例来看多个拦截器的拦截流程。

【实例 14.3】模拟添加用户准入许可的拦截器。只允许 access 属性为 1 的用户访问网站主页面。

1. 编写拦截器实现类 com.inspur.interceptor. PermissionInterceptor

新添加的拦截器对会话中保存的用户访问许可信息进行过滤，不允许 access 为 0 的用户通过检查。

```
public class PermissionInterceptor implements HandlerInterceptor{
    /**
     * 预处理回调方法
```

```
     */
    public boolean preHandle(HttpServletRequest request, HttpServletResponse
         response, Object handler) throws Exception {
         String access = (String) request.getSession().getAttribute("useraccess");
         if(access.equals("1")){
              System.out.println("2.LoggingInterceptor 的 preHandle 方法正在被执
         行……[请求通过拦截器检查]");
              return true;
         }else{
              System.out.println("2.LoggingInterceptor 的 preHandle 方法正在被执
         行……[请求被拦截]");
              request.setAttribute("msg", "用户不合法，请求被第二个拦截器拦截！");
              request.getRequestDispatcher("/login").forward(request, response);
              return false;
         }
    }
    /**
     * 后处理回调方法
     */
    public void postHandle(HttpServletRequest request, HttpServletResponse
        response,Object handler, ModelAndView mv) throws Exception {
        System.out.println("2.LoggingInterceptor 的 postHandle 方法正在被执行……");
    }
    /**
     * 请求处理完毕回调方法
     */
    public void afterCompletion(HttpServletRequest request,
        HttpServletResponse response, Object hanler, Exception exp)
        throws Exception {
        System.out.println("2.LoggingInterceptor 的 afterCompletion 方法正在被执
    行……");
    }
}
```

2. 在前端控制器中添加新拦截器配置

新添加的拦截器配置在原有拦截器后面，对除 login 之外的所有请求都有效。

```
<mvc:interceptors>
   <mvc:interceptor>
      <!-- 配置拦截器生效的路径 -->
      <mvc:mapping path="/**"/>
      <!-- 配置拦截器不生效的路径，在生效的路径范围内，消除不生效的路径 -->
      <mvc:exclude-mapping path="/*login"/>
      <!-- 拦截器的位置 -->
      <bean class="com.inspur.interceptor.LoginInterceptor"></bean>
   </mvc:interceptor>
   <mvc:interceptor>
      <mvc:mapping path="/**"/>
      <mvc:exclude-mapping path="/*login"/>
      <bean class="com.inspur.interceptor. PermissionInterceptor"></bean>
```

```
        </mvc:interceptor>
    </mvc:interceptors>
```

3．执行测试

（1）正常登录访问网站主界面测试

在浏览器中访问地址 http://127.0.0.1:8080/mvc-interceptor/login，输入用户名（张三）及密码（123）进行登录。使当前会话中保存张三的登录状态，然后进行如下测试。

请求地址：http://127.0.0.1:8080/mvc-interceptor/gomain。

页面显示如图 14-9 所示。

```
浪潮集团
欢迎您：张三
```

图 14-9　多个拦截器的正常执行流程的页面显示

控制台显示如图 14-10 所示。

```
1.LoginInterceptor的preHandle方法正在被执行...[请求通过拦截器检查]
2.LoggingInterceptor的preHandle方法正在被执行...[请求通过拦截器检查]
控制器方法正在执行...
2.LoggingInterceptor的postHandle方法正在被执行...
1.LoginInterceptor的postHandle方法正在被执行...
网站主页面加载成功...
2.LoggingInterceptor的afterCompletion方法正在被执行...
1.LoginInterceptor的afterCompletion方法正在被执行...
```

图 14-10　多个拦截器的正常执行流程的控制台显示

（2）请求被第二个拦截器中断测试

由于第二个拦截器中设定不允许 id 为 2 的用户访问系统，使用用户名为李四、密码为 123456 进行网站登录，使当前会话中保存李四的登录状态。然后进行如下测试。

请求地址：http://127.0.0.1:8080/mvc-interceptor/gomain。

页面显示如图 14-11 所示。

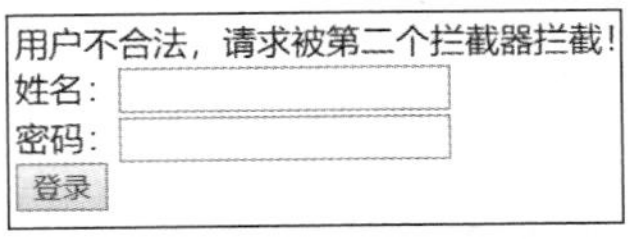

```
用户不合法，请求被第二个拦截器拦截!
姓名：
密码：
登录
```

图 14-11　多个拦截器的中断执行流程的页面显示

控制台显示如图 14-12 所示。

```
1.LoginInterceptor的preHandle方法正在被执行...[请求通过拦截器检查]
2.LoggingInterceptor的preHandle方法正在被执行...[请求被拦截]
1.LoginInterceptor的afterCompletion方法正在被执行...
```

图 14-12　多个拦截器的中断执行流程控制台显示

根据以上两个案例可以看出，多个拦截器的执行流程如图 14-13 所示。当流程不被任意一个拦截器拦截时，会按照前端控制器中的配置顺序，首先执行所有的预处理方法，然后执行目标方法，目标方法执行完后，再按照配置顺序倒序执行后处理方法。接下来进行视图渲染，视图渲染完毕，再按照配置顺序倒序执行请求处理完毕的回调方法。

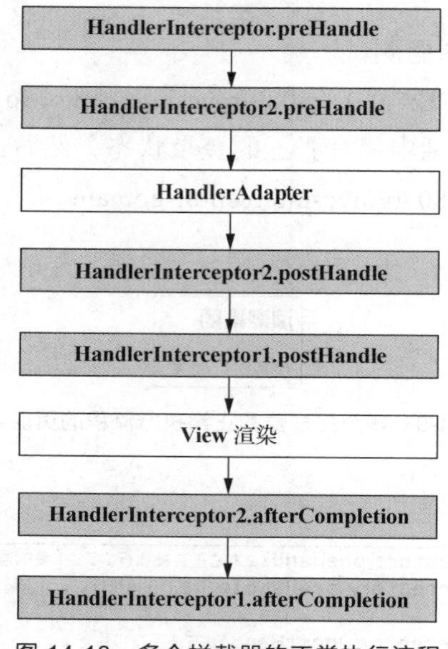

图 14-13　多个拦截器的正常执行流程

请求被第二个拦截器拦截后，会中断请求的处理，转而执行第一个拦截器的请求完毕处理方法，如图 14-14 所示。

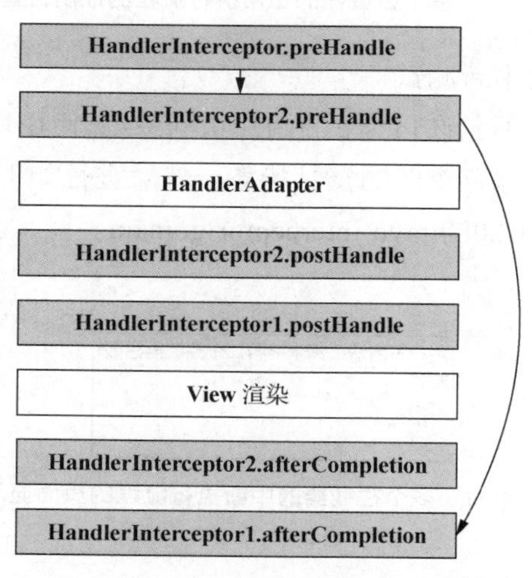

图 14-14　多个拦截器的中断执行流程

14.3.3　性能监控

接下来介绍一个拦截器常见的应用：性能监控。软件的运行效率是评判一个软件好坏的重要标准。在业务进行过程中可能需要对请求的处理过程进行监控，记录一下请求的处理时间，得到一些慢请求（如处理时间超过 500ms），从而进行性能改进，一般的反向代理服务器（如 apache）都有这个

性能监控

功能，此处演示使用拦截器怎么实现性能监控。

首先对业务需求进行分析。

（1）在请求发出后、进入处理器之前开始计时，需要借助拦截器的 preHandle 记录开始时间。

（2）在请求处理完毕后记录结束时间，并用结束时间减去开始时间得到这次请求的处理时间，这部分可以借助拦截器的 afterCompletion 实现。

但还存在一个问题：不管用户请求多少次都只有一个拦截器实现，那应该怎么记录每个请求的处理时间呢？

解决方案是使用 ThreadLocal，它是线程绑定的变量，提供线程局部变量（一个线程一个 ThreadLocal，A 线程的 ThreadLocal 只能看到 A 线程的 ThreadLocal，不能看到 B 线程的 ThreadLocal）。

然后定义一个拦截器，由于该拦截器中用到了 preHandle 和 afterCompletion 两个方法，所以可以通过 HandlerInterceptorAdapter 适配器来实现。

```
public class StopWatchHandlerInterceptor extends HandlerInterceptorAdapter{
    private NamedThreadLocal<Long> startTimeThreadLocal = new
    NamedThreadLocal<Long>("StopWatch-StartTime");
    /**
     * 请求发出后，执行目标方法之前记录当前请求处理开始时间
     */
    public boolean preHandle(HttpServletRequest request, HttpServletResponse
        response, Object handler)throws Exception {

        long beginTime = System.currentTimeMillis();// 获取业务开始时间
        startTimeThreadLocal.set(beginTime);// 将请求处理开始时间绑定到线程变量
（该数据仅当前线程可见）
        return true;
    }
    /**
     * 请求处理完毕后，记录请求处理消耗时间，如果处理时间超过 500ms 则认为是慢请求，需
要记录日志
     */
    public void afterCompletion(HttpServletRequest request, HttpServlet
        Response response, Object handler, Exception ex)throws Exception {
        long endTime = System.currentTimeMillis();// 获取业务结束时间
        long beginTime = startTimeThreadLocal.get();// 得到线程绑定的局部变量（开
始时间）
        // 计算消耗时间，此处认为处理时间超过 500ms 的请求为慢请求
        long consumeTime = endTime - beginTime;
        if (consumeTime > 500) {
        System.out.println(request.getRequestURI()+"请求处理耗时超过 500ms，请
关注！");
        }
        System.out.println("线程计时结束:"+String.format("%s consume %d millis",
request.getRequestURI(), consumeTime));
```

```
        }
    }
```

在前端控制器中进行拦截器的配置，由于该拦截器需要监控所有请求的处理效率，所以配置该拦截器对所有请求有效。

```
<!-- 配置拦截器 -->
    <mvc:interceptors>
        <bean class="com.inspur.interceptor.HelloInterceptor"></bean>
        <bean class="com.inspur.interceptor.StopWatchHandlerInterceptor"></bean>
        <mvc:interceptor>
            <!-- 配置拦截器生效的路径 -->
            <mvc:mapping path="/**"/>
            <!-- 配置拦截器不生效的路径，在生效的路径范围内，消除不生效的路径 -->
            <mvc:exclude-mapping path="/*login"/>
            <!-- 拦截器的位置 -->
            <bean class="com.inspur.interceptor.LoginInterceptor"></bean>
        </mvc:interceptor>
        <mvc:interceptor>
            <mvc:mapping path="/**"/>
            <mvc:exclude-mapping path="/*login"/>
        <bean class="com.inspur.interceptor.PermissionInterceptor"></bean>
        </mvc:interceptor>
    </mvc:interceptors>
```

接下来对性能监控拦截器进行测试。

请求地址：http://127.0.0.1:8080/mvc-interceptor/login。

页面显示如图 14-15 所示。

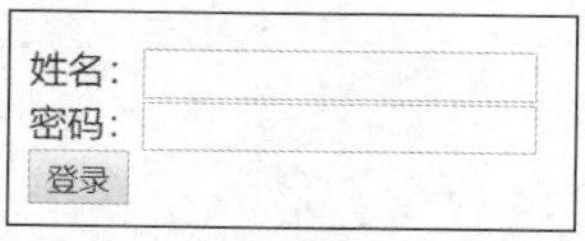

姓名：

密码：

登录

图 14-15　性能监控页面显示

控制台显示如图 14-16 所示。

```
线程计时结束：/mvc-interceptor/login consume 52 millis
```

图 14-16　性能监控控制台显示

本章小结

本章介绍了拦截器的概念及常见应用，拦截器的配置方式，多个拦截的执行流程。读者需要掌握 Spring MVC 框架中的拦截器接口 HandlerInterceptor 和拦截器适配器 HandlerInterceptorAdapter，还需要通过用户登录状态拦截器的案例掌握拦截器的配置方法，多个拦截器的执行流程，以及理解并掌握性能监控案例。

习题

1. 简述 Spring MVC 中拦截器的定义方式。
2. 简述多个拦截器的正常流程和中断流程图。

上机指导

开发一个 Java Web 工程，至少实现用户登录模块，并添加用户访问量拦截器，记录网站主页面的用户访问量。

15 第15章　Spring MVC对Ajax的支持

学习目标

- 理解 Ajax 的概念
- 理解 JSON 的概念
- 理解直接的 Ajax 处理的方法
- 掌握通过注解进行 Ajax 处理的方法
- 掌握使用 ResponseEntity 支持 Ajax 的方法
- 掌握对 Ajax 返回 XML 的支持的方法
- 理解 HttpMessageConverter 机制
- 了解 Ajax 请求过程中的内容协商

15.1　Ajax 简介

Ajax 简介

Web 1.0 以静态、单向阅读为主，进入 Web 2.0 时代后，新的交互方式使网络的用户体验更加丰富，用户的操作可以在不刷新窗口的前提下完成。

如果说 Web 2.0 以人为本，面向未来，开创了新的时代，那么 Ajax（Asynchronous JavaScript And XML，异步的 JavaScript 和 XML）就是推动这一时代进步的“助推器”。图 15-1 所示的是 Web 2.0 的常见应用。

使用 Ajax 可以不刷新整个页面，只刷新局部。它的优点如下。

① 只更新部分页面，可以有效利用带宽（见图 15-2）。

② 提供连续的用户体验（见图 15-3）。

③ 提供类似 C/S 的交互效果（见图 15-4），操作更方便。

那什么是 Ajax 呢？Ajax 是指一种创建交互式网页应用的网页开发技术，尤其是在无须重新加载整个网页的情况下，能够更新部分网页。Ajax 的应用实例有新浪微博、地图、百度搜索等。Ajax 的 Logo 及主要涉及的技术如图 15-5 所示。

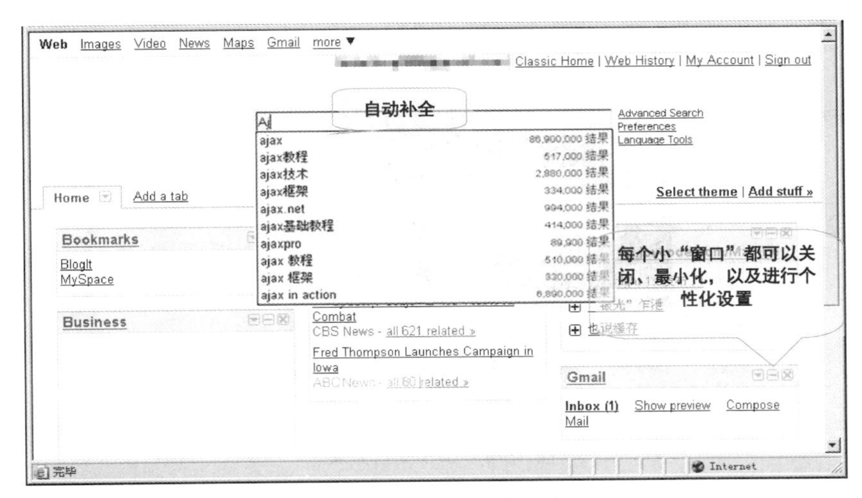

图 15-1　Web 2.0 的常见应用

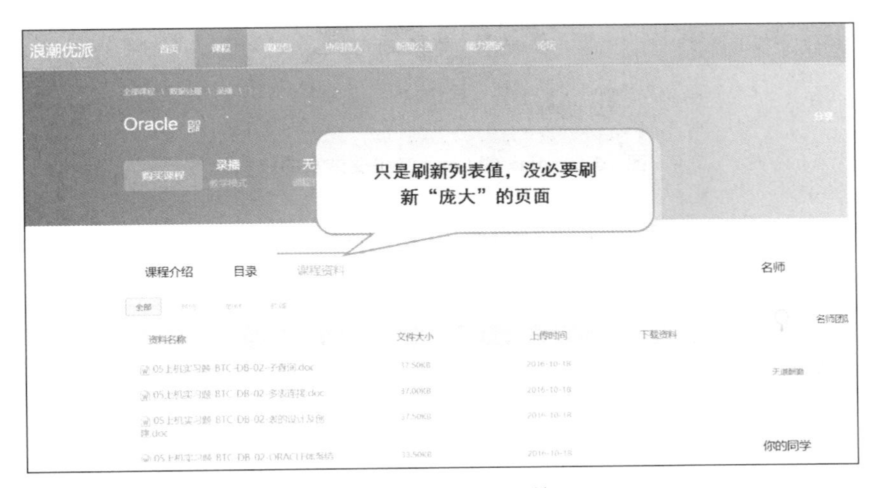

图 15-2　Ajax 局部刷新

图 15-3　连续的用户体验

图 15-4　C/S 交互效果

图 15-5　Ajax 的 Logo 及主要涉及的技术

Ajax 的 Logo 中每个字母对应了一个单词，具体介绍如下。

① Asynchronous（异步的）：Ajax 采用异步模式，表示发送请求后不等返回结果，由回调函数处理结果。

② JavaScript：Ajax 使用 JavaScript 作为开发 Web 页面的脚本语言，用于向服务器发起请求，获得返回结果，更新页面。

③ And：是连接词。

④ XML：可扩展标记语言，Ajax 使用其封装数据进行传输。

在 Spring MVC 的实际应用中可以进行数据验证、数据获取等操作，那么意味着在 Spring MVC 中引入 Ajax 就可以实现异步更新，这也是本章要介绍的内容。

15.2　JSON 简介

JSON 简介

在 Spring MVC 和 Ajax 进行动态交互的过程中，涉及交互数据的绑定，那么首先来认识一下 JSON。JSON（JavaScript Object Notation，JavaScript 对象标记）是一种非常流行的数据格式。它与 XML 非常相似，都是用来存储数据的，但是 JSON 相对于 XML 来说解析速度更快、占用空间更小。在实际应用当中，使用 JSON 格式的数据进行前后台的数据交互是很常见的。

JSON 是一种轻量级的数据交互格式。它基于 JavaScript 的一个子集，采用一种独立于编程语言的文本格式来存储和表示数据，易于阅读和编写，同时也易于机器解析和生成，这就使得 JSON 成为一种理想的数据交互语言，并得到了广泛应用，绝大多数开发语言都有处理 JSON

的 API。

JSON 是基于纯文本的数据格式，使用 JSON 可以传输一个简单的 String、Number 类型的数据，也可以传输一个数组或一个复杂的 JavaBean 对象。

JSON 有以下两种数据结构。

1. 对象结构

对象是一个无序的“名/值”对集合。一个对象以“{”开始、以“}”结束。每个“名称”后跟一个“:”；“名/值”对之间使用“,”分隔，其存储形式如图 15-6 所示。

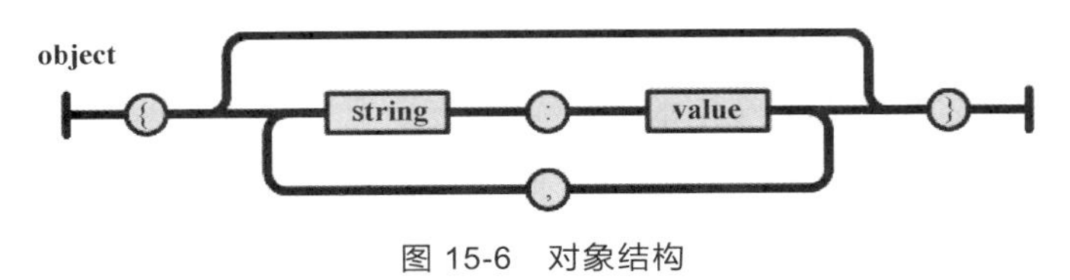

图 15-6　对象结构

对象结构的语法结构如下。

```
{
    key1:value1,
    key2:value2,
    …
}
```

其中，关键字（Key）必须为 String 类型，值（Value）可以是 String、Number、Object 等数据类型。例如，一个 address 对象包含城市、街道、邮编等信息，使用 JSON 的表示形式如下：

```
{"city":"jinan","street":"jingshilu","postcode":250000}
```

2. 数组结构

数组结构以“[”开始、以“]”结束。中间部分由 0 个或多个以英文“,”分隔的值的列表组成，其存储形式如图 15-7 所示。

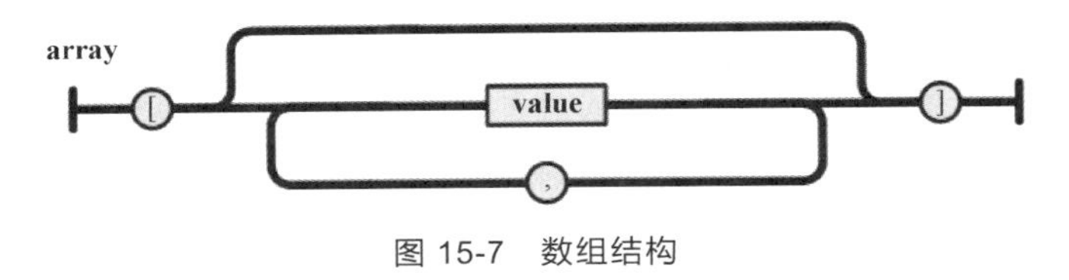

图 15-7　数组结构

数组结构的语法结构如下。

```
[
    value1,
    value2,
    …
]
```

例如，一个数组包含了 String、Number、Boolean 类型数据，使用 JSON 的表示形式如下：

```
["abc",123,false]
```

上面两种（对象、数组）数据结构也可以分别组合成更为复杂的数据结构。例如，一个 person 对象包含 name、hobby 和 address 对象，其代码表现形式如下。

```
{
    "name":"zhangsan",
    "hobby":["Basketball","Table tennis","Swimming"],
    "address":{
        "city":"jinan",
        "street":"jingshilu",
        "postcode":250000
    }
}
```

需要注意的是，如果使用 JSON 存储单个数据（如“abc”），一定要使用数组结构，不要使用对象结构，因为对象结构必须是“名称:值”的形式。

JSON 在 Spring MVC 中是怎么应用的呢？

为了实现浏览器与控制器类（Controller）之间的数据交互，Spring MVC 提供了一个 HttpMessageConverter<T>接口来完成此项工作。该接口主要用于将请求信息中的数据转换为一个类型为 T 的对象，并将此类型为 T 的对象绑定到请求方法的参数中，或者将对象转换为响应信息传递给浏览器显示。

Spring MVC 为 HttpMessageConverter<T>接口提供了很多实现类，这些实现类可以对不同类型的数据进行信息转换。其中 MappingJackson2HttpMessageConverter 是 Spring MVC 默认处理 JSON 格式请求响应的实现类。该实现类利用 Jackson 开源包读写 JSON 数据，将 Java 对象转换为 JSON 对象和 XML 文档，同时也可以将 JSON 对象和 XML 文档转换为 Java 对象。

要使用 MappingJackson2HttpMessageConverter 进行数据转换，就要用到 Jackson 的开源包，Jackson 是解析 JSON 和 XML 的一个框架，Java 生态圈中有很多处理 JSON 和 XML 格式化的类库，Jackson 是其中比较著名的一个。本章所要使用的 Jackson 的开源包具体如下：jackson-annotations-2.9.8.jar、jackson-core-2.9.8.jar、jackson-databind-2.9.8.jar。

以上 3 个 Jackson 的开源包读者可以通过官方网站下载。

15.3 直接的 Ajax 处理

在 Spring MVC 中可以直接使用实体对象接收来自前台的 JSON 格式数据。返回页面的 JSON 类型结果只要在 Controller 的方法里面，直接使用 response 输出要返回的 Ajax 数据，然后 return null 就可以了。示例如下：其中 Userssm 是用来接收页面数据的 POJO 类，包含 id、username 两个属性的声明和 get/set 方法。

直接的 Ajax 处理

1. Controller 示例

```
@RequestMapping(value = "/hello1")
public ModelAndView handleRequest(Userssm um,HttpServletResponse response)
    throws IOException {
    response.setCharacterEncoding("utf-8");
    response.getWriter().println("{\"id\":\""+um.getId()+"\", \"username\":\
        ""+um.getUsername()+"\"}");
```

```
        return null;
    }
```

2. 客户端示例

```
<%@ page language="java" contentType="text/html; charset=UTF-8 " pageEncoding=
  "UTF-8"%>
<script language="javascript" src="${ContextPath}/js/jquery-2.1.1.min.js">
</script>
<script language="javascript">
  $().ready(function(){
  $.getJSON('hello1',{id:'1',username:'test'},
      function(data){
         alert(data.id+" , "+data.username);
      });
    });
</script>
</html>
```

3. 进行测试

由于在 web.xml 中配置的“/”会将页面中引入的静态文件也进行拦截，而拦截后页面中将找不到这些静态资源文件，这样就会引起页面报错。从而需要在 Spring 配置文件中增加静态资源的访问映射配置。

```
<mvc:resources location="/js/" mapping="/js/**"></mvc:resources>
```

之后就可以启动服务器进行测试了。运行结果如图 15-8 所示。

图 15-8　直接的 Ajax 处理结果

15.4　通过注解进行的 Ajax 处理

通过注解进行的 Ajax 处理

除了直接的 Ajax 处理方式，还可以使用注解的方式进行 JSON 数据交换。在 Controller 中解析 JSON 数据和从 Controller 中返回 JSON 数据使用的注解分别是@RequestBody 和@ResponseBody。具体说明如表 15-1 所示。

表 15-1　　JSON 注解说明

注解	说明
@RequestBody	用于将请求体中的数据绑定到方法的形参中
@ResponseBody	用于 return 返回对象上

下面来详细介绍这两个注解。

15.4.1 @RequestBody

@RequestBody 注解主要用于读取 HTTP 请求的内容（字符串），并将之转换成对应的 Java 对象，同时绑定到 Controller 方法的参数上。书写形式如下。

```
@RequestMapping(value = "/hello")
public String handleRequest(@RequestBody String body)
```

为了进一步理解 Spring MVC 中 JSON 数据的交互，接下来通过一个案例演示具体的实现过程。

（1）项目准备

首先，为了使用 JSON，需要导入 Jackson 相关的 JAR 包，如图 15-9 所示。

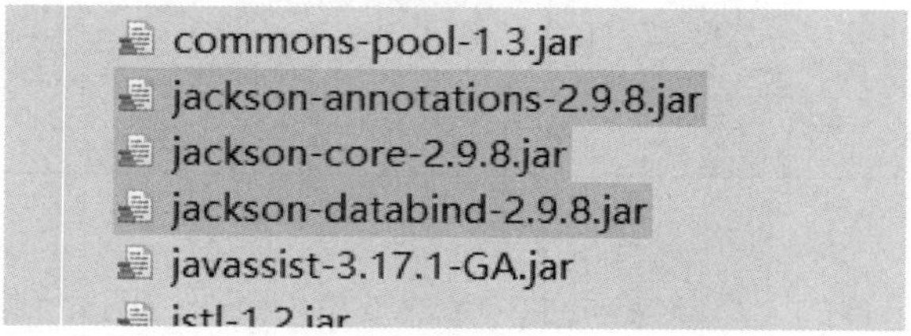

commons-pool-1.3.jar
jackson-annotations-2.9.8.jar
jackson-core-2.9.8.jar
jackson-databind-2.9.8.jar
javassist-3.17.1-GA.jar
jstl-1.2.jar

图 15-9　导入 Jackson 相关的 JAR 包

（2）添加配置项

要配置 JSON 转换器，可使用<mvc:annotation-driven/>进行配置，它会自动注册 RequestMappingHandlerAdapter 和 RequestMappingHandlerMapping 两个 Bean，并提供对读写 XML 和 JSON 等功能的支持，不需要添加其他配置信息。

（3）Controller 书写

```
/**
  * 使用@RequestBody 将 JSON 数据转换成 String 类型的数据
  * @param reqBody
  * @param response
  * @return
  */
 @RequestMapping(value = "/hello2")
 public ModelAndView handleRequest(@RequestBody String reqBody,HttpServlet
     Response response){
     response.setCharacterEncoding("utf-8");
     try {
          response.getWriter().println(reqBody);
     } catch (IOException e) {
          // TODO Auto-generated catch block
          e.printStackTrace();
     }
     return null;
 }
```

（4）前台 Ajax 请求

```
<%@ page language="java" contentType="text/html; charset=UTF-8" page Encoding=
    "UTF-8"%>
<script type="text/javascript" src="../js/jquery-2.1.1.min.js"></script>
<script>
```

```
$().ready(function(){
        $.ajax({
           url : '../hello2',
           type : 'post',
           data : JSON.stringify({id:"1",username:"test"}),
           contentType : 'application/json; charset=utf-8',
               success:function(data){
                  alert(data);
               }
        });
});
</script>
```

在上述 JSON 测试表单中，当页面加载的时候会执行页面的准备函数，在函数中使用了 jQuery 的 Ajax 方式将 JSON 格式的数据传递到以“/hello2”结尾的请求中。

需要注意以下几点。

- 需要添加 jQuery 依赖文件。
- data：即请求时携带的数据，当使用 JSON 格式时，要注意编写规范。
- contentType：当请求数据为 JSON 格式时，值必须为 application/json。

测试结果如图 15-10 所示。

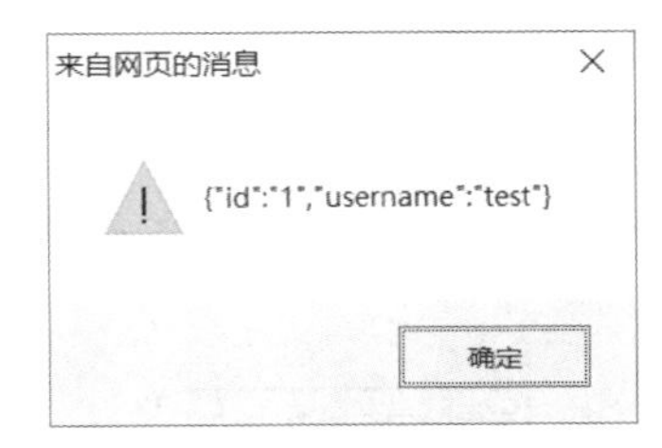

图 15-10　将 JSON 数据转换为 String 类型的数据

@RequestBody 除了可以将 JSON 数据转换为 String 类型外，还可以将其转换为 List 甚至是 Bean 等 Java 对象。示例如下，其中 Userssm um 是 Bean 对象，具体应用方式将在 15.4.2 节详细叙述。

```
@RequestMapping(value = "/hello")
public String handleRequest(@RequestBody  Userssm um)
```

15.4.2 @ResponseBody

@ResponseBody 用于将返回值以设定的返回类型响应至页面，示例如下，将 String 类型返回至页面进行解析。

```
@RequestMapping(value = "/hello")
@ResponseBody
public String handleRequest(@RequestBody String body)
```

可以使用@RequestBody 自动获取 Ajax 上传的数据，同时也可以使用@ResponseBody 把要返回的对象自动拼成 JSON 的格式返回。配置依旧使用<mvc:annotation-driven/>方式即可。Controller 书写方法如下（同时论证使用@RequestBody 将 JSON 数据转换成 Bean 对象）。

```
/**
```

```
  * 将 JSON 数据转为 Bean 对象，再返回 Bean 对象至页面
  * @param um
  * @return
  */
@RequestMapping(value = "/hello3")
@ResponseBody
public Userssm handleRequest(@RequestBody Userssm um) {
    return um;
}
```

前台 Ajax 请求如下。

```
<%@ page language="java" contentType="text/html; charset=UTF-8" pageEncoding=
    "UTF-8"%>
<script type="text/javascript" src="../js/jquery-2.1.1.min.js"></script>
<script>
$().ready(function(){
     $.ajax({
             url : '../hello3',
             type : 'post',
             data : JSON.stringify({id:"1",username:"test"}),
             contentType : 'application/json; charset=utf-8',
             success:function(data){
                     alert(data.id+"  "+data.username);
             }
     });
});
</script>
```

测试结果如图 15-11 所示。

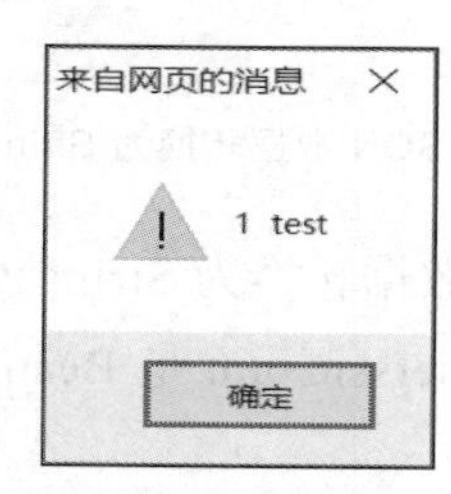

图 15-11 @ResponseBody 返回数据至页面

通过测试可以看到 Controller 的方法返回一个 Bean 对象，但是@ResponseBody 会把这个对象自动变成 JSON 格式，再传回客户端，非常方便。

当然，@ResponseBody 也支持集合对象自动变成 JSON 格式，可修改 Controller 方法如下。

```
/**
    * @ResponseBody 返回集合数据
    * @param um
    * @return
    */
   @RequestMapping(value = "/hello4")
   @ResponseBody
   public List<Userssm> handleRequestHello4(@RequestBody Userssm um) {
       List<Userssm> list = new ArrayList<Userssm>();
       list.add(um);
```

```
        Userssm um2 = new Userssm();
        um2.setId("22");
        um2.setUsername("222");
        list.add(um2);
        return list;
    }
```

测试使用的页面代码。

```
<%@ page language="java" contentType="text/html; charset=UTF-8" pageEncoding=
    "UTF-8"%>
<script type="text/javascript" src="../js/jquery-2.1.1.min.js"></script>
<script>
$().ready(function(){
     $.ajax({
            url : '../hello4',
            type : 'post',
            data : JSON.stringify({id:"1",username:"test"}),
            contentType : 'application/json; charset=utf-8',
            success:function(data){
                $.each(data,function(index,v){
                    alert("id="+v.id+",name="+v.username);
                });
            }
     });
});
</script>
```

显示结果如图 15-12 所示。

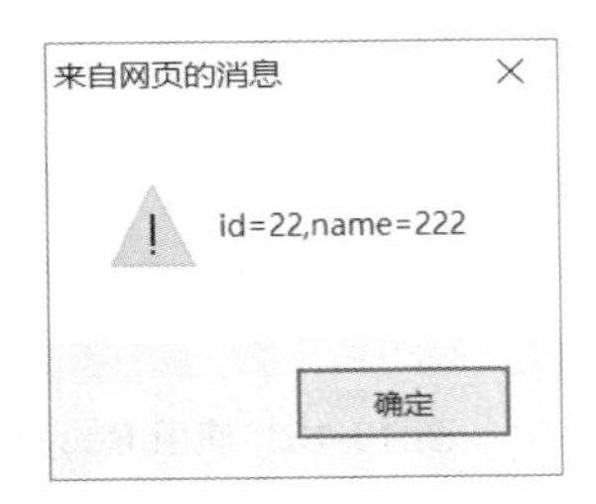

图 15-12　使用@ResponseBody 返回集合数据

@RequestBody/@ResponseBody 只能访问到报文体，不能访问到报文头。

15.5　使用 ResponseEntity 支持 Ajax

使用 Response-Entity 支持 Ajax

使用 Spring 时，要达到同一目的通常有多种方法，处理 HTTP 响应也是一样的。@RequestBody/@ResponseBody 可以在请求的过程中进行数据传递，但不能访问到报文头，而 ResponseEntity 却可以。接下来介绍如何通过 ResponseEntity 设置 HTTP 报文头信息。ResponseEntity 是在 org.springframework.http.HttpEntity 的基础上添加了 http status code（HTTP 状态码），用来设定返回状态。

其主要的变化在 Controller 的方法上，页面不用变化，具体代码如下。

```
/**
  * 使用 HttpEntity/ResponseEntity 来支持 Ajax
  * @param req
  * @param um
  * @return
  */
@RequestMapping(value = "/hello5")
public ResponseEntity<List<Userssm>> handleRequest(HttpEntity<String> req,
   Userssm um) {
   System.out.println("req headers="+req.getHeaders()+", reqBody="+req. getBody());
   um.setUsername(um.getUsername()+",server");
   List<Userssm> list = new ArrayList<Userssm>();
   list.add(um);
   Userssm um2 = new Userssm();
   um2.setId("22");
   um2.setUsername("222");
   list.add(um2);
   ResponseEntity<List<Userssm>> ret = new ResponseEntity<List<Userssm>> (list,
   HttpStatus.OK);
   return ret;
}
```

测试结果如图 15-13 所示。

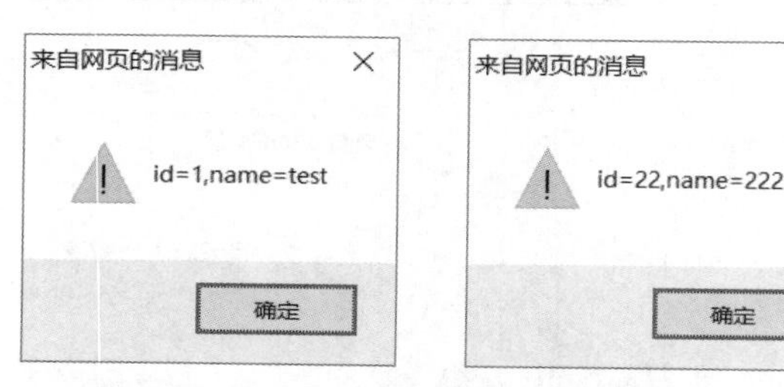

图 15-13 使用 ResponseEntity 设置响应头信息

15.6 对 Ajax 返回 XML 的支持

对 Ajax 返回 XML 的支持

前面的 Ajax 使用的是 JSON 格式，在 Spring MVC 与 Ajax 的数据交互过程中还有一种常见的数据格式就是 XML，下面介绍 Spring MVC 对 XML 的支持。

要让 Spring MVC 支持 XML 格式，需要加入 JAXB（Java Architecture for XML Binding，XML 绑定的 Java 体系结构）支持。JAXB 是一个业界的标准，是一项可以根据 XML Schema 产生 Java 类的技术。该过程中，JAXB 也提供了将 XML 实例文档反向生成 Java 对象树的方法，并能将 Java 对象树的内容重新写到 XML 实例文档中。支持 JAXB 需要加入的 JAR 包为 jaxb-api-2.3.1.jar 和 jaxb-impl-2.1.jar。

在要返回的 Bean 对象头上使用如下注解：@XmlRootElement(name = "testxml")。

① 此注解表明返回的 XML 的根元素名称为 testxml，由于 XML 是单根的，所以只能返回一个对象，而不能返回一个 list，如果要返回多个值，可以在这个对象中包含多个其他对象。

② 返回的结果同样用@ResponseBody 注解即可，这个注解会根据请求的类型自动决定返回 JSON 还是 XML。默认返回的是 JSON 格式，如果要返回 XML 格式，那么在请求的时候，就要指定 accept=application/xml。

（1）Controller 方法

```
@RequestMapping(value = "/hello6")
@ResponseBody
public Userssm handleRequestHello6(HttpEntity<String> req, Userssm um) {
    PhoneNumber pnm = new PhoneNumber("123","321");
    PhoneNumber pnm2 = new PhoneNumber("2222","333");
    List<PhoneNumber> tempList = new ArrayList<PhoneNumber>();
    tempList.add(pnm2);
    tempList.add(pnm);
    um.setPm(tempList);
    return um;
}
```

（2）Bean 对象

为了增加 XML 数据层层嵌套的效果，可在 Userssm 对象中添加电话号码属性对象 PhoneNumber，具体代码如下所示。

```
package com.inspur.ssm.pojo;
public class PhoneNumber {
    private String areaCode;
    private String phoneNumber;
    public String getAreaCode() {
        return areaCode;
    }
    public void setAreaCode(String areaCode) {
        this.areaCode = areaCode;
    }
    public String getPhoneNumber() {
        return phoneNumber;
    }
    public void setPhoneNumber(String phoneNumber) {
        this.phoneNumber = phoneNumber;
    }
    public PhoneNumber(String areaCode, String phoneNumber) {

        this.areaCode = areaCode;
        this.phoneNumber = phoneNumber;
    }
    public PhoneNumber() {

    }
}
```

Userssm 对象代码如下所示。

```
package com.inspur.ssm.pojo;
import java.util.List;
import javax.xml.bind.annotation.XmlRootElement;
@XmlRootElement(name = "testxml")
public class Userssm {
     private String id;
     private String username;
     private List<PhoneNumber> pm ;
     public String getId() {
         return id;
     }
     public void setId(String id) {
         this.id = id;
     }
     public String getUsername() {
         return username;
     }
     public void setUsername(String username) {
         this.username = username;
     }
     public List<PhoneNumber> getPm() {
         return pm;
     }
     public void setPm(List<PhoneNumber> pm) {
         this.pm = pm;
     }
}
```

（3）示例的页面

```
<%@ page language="java" contentType="text/html; charset=UTF-8" pageEncoding=
     "UTF-8"%>
<script type="text/javascript" src="../js/jquery-2.1.1.min.js"></script>
<script>
$().ready(function(){
     $.ajax({
             url : '../hello6',
             type : 'post',
             data : JSON.stringify({id:"1",username:"test"}),
             contentType : 'application/xml; charset=utf-8',
             error: function(){ alert('Error loading XMLdocument'); },
             success: function(xml){
                   $(xml).find("testxml").children("pm").each(function(i){
                      var areaCode=$(this).children("areaCode").text();
                      var phoneNumber=$(this).children("phoneNumber").text();
                   alert("areaCode="+areaCode+",phoneNumber="+phoneNumber);
                   });
             }
     });
});
</script>
```

（4）返回的 XML

```
<?xml version="1.0" encoding="UTF-8" standalone="yes"?>
<testxml>
    <pm>
```

```
        <areaCode>2222</areaCode>
        <phoneNumber>333</phoneNumber>
    </pm>
    <pm>
        <areaCode>123</areaCode>
        <phoneNumber>321</phoneNumber>
    </pm>
</testxml>
```

15.7　HttpMessageConverter

前面的程序实现对 JSON 和 XML 的支持之所以能正常运行，“幕后英雄”就是 HttpMessageConverter，它负责对 HTTP 传入和传出的内容进行格式转换。如@RequestBody 是将 HTTP 请求正文插入方法中，它就会使用适合的 HttpMessageConverter 将请求体写入某个对象。而@ResponseBody 将内容或对象作为 HTTP 响应正文返回，使用@ResponseBody 将会跳过视图处理部分，直接调用合适的 HttpMessageConverter，并将返回值写入输出流。只要开启<mvc:annotation-driven/>，它就会给 AnnotationMethodHandlerAdapter 初始化 7 个转换器，可以通过调用 AnnotationMethodHandlerAdapter 类的 getMessageConverts 方法来获取转换器的一个集合 List<HttpMessageConverter>，默认开启的有 ByteArrayHttpMessageConverter、StringHttpMessageConverter、ResourceHttpMessageConverter、SourceHttpMessageConverter<T>、XmlAwareFormHttpMessageConverter、Jaxb2RootElementHttpMessageConverter、MappingJacksonHttpMessageConverter。

Spring 是如何寻找最佳的 HttpMessageConverter 的呢？

最基本的方式就是通过请求的 Accept 里面的格式来匹配，如有 accept=application/json，就使用 JSON 的 HttpMessageConverter；如有 accept=application/xml，就使用 XML 的 HttpMessageConverter。

Spring 4.x 以后，MappingJacksonHttpMessageConverter 更改为 MappingJackson2HttpMessageConverter。

15.8　Ajax 请求过程中的内容协商

什么是内容协商？简单来说，就是同一资源可以有多种表现形式，如 XML、JSON 等，具体使用哪种表现形式是可以协商的。这是 RESTful 的一个重要特性，Spring MVC 也支持这个功能。

Ajax 请求过程中的内容协商

Spring MVC REST 是如何决定采用何种方式（视图）来展示内容的呢？

（1）根据 HTTP 请求的 Header 中的 Accept 属性的值来判读。

① Accept: application/xml 将返回 XML 格式的数据。

② Accept: application/json 将返回 JSON 格式的数据。

图 15-14 所示是在请求的时候设置的数据格式。

```
<%@ page language="java" contentType="text/html; charset=UTF-8" pageEncoding="UTF-8"%>
<script type="text/javascript" src="../js/jquery-2.1.1.min.js"></script>
<script>
$().ready(function(){
    $.ajax({
            url : '../hello6',
            type : 'post',
            data : JSON.stringify({id:"1",username:"test"}),
            contentType : 'application/xml; charset=utf-8',
            error: function(){ alert('Error loading XMLdocument'); },
            success: function(xml){
                $(xml).find("testxml").children("pm").each(function(i){
                    var areaCode=$(this).children("areaCode").text();
                    var phoneNumber=$(this).children("phoneNumber").text();
                    alert("areaCode="+areaCode+",phoneNumber="+phoneNumber);
            });
        }
    });
});
</script>
```

图 15-14 请求数据格式设置

项目正常运行后，可通过浏览器（这里使用的是 Chrome 浏览器）的开发模式看到图 15-15 所示的请求头部信息。

```
Status Code: ● 200
Remote Address: [::1]:8080
Referrer Policy: no-referrer-when-downgrade
▸ Response Headers (3)
▾ Request Headers    view source
Accept: application/xml, text/xml, */*; q=0.01
Accept-Encoding: gzip, deflate, br
Accept-Language: zh-CN,zh;q=0.9
```

图 15-15 请求头部信息

根据 Header 中的 Accept 属性值来判读何种展示方式的优缺点如下。

优点：这种方式是理想的标准方式。

缺点：由于浏览器的差异，导致发送的 Accept Header 可能会不一样，从而导致服务器不知要返回什么格式的数据。

（2）根据扩展名来判断。

如图 15-16 所示，在访问地址后加上后缀扩展名。

① /test.xml 将返回 XML 格式的数据。

② /test.json 将返回 JSON 格式的数据。

③ /test.html 将返回 HTML 格式的数据。

```
1 <%@ page language="java" contentType="text/html; charset=UTF-8" pageEncoding="UTF-8"%>
2 <script type="text/javascript" src="../js/jquery-2.1.1.min.js"></script>
3 <script>
4 $().ready(function(){
5     $.ajax({
6         url : '../hello6.xml',
7         type : 'post',
8         data : JSON.stringify({id:"1",username:"test"}),
9 //        contentType : 'application/xml; charset=utf-8',
10        error: function(){ alert('Error loading XMLdocument'); },
11        success: function(xml){
12            $(xml).find("testxml").children("pm").each(function(i){
13                var areaCode=$(this).children("areaCode").text();
14                var phoneNumber=$(this).children("phoneNumber").text();
15                alert("areaCode="+areaCode+",phoneNumber="+phoneNumber);
16            });
17        }
18    });
19 });
20 </script>
```

图 15-16　访问地址后加上后缀扩展名

根据扩展名来判断的优缺点如下。

优点：在实际环境中使用还是较多的，因为这种方式更符合程序员的习惯。

缺点：丧失了同一 URL 的多种展现方式。

（3）根据参数来判断。

① /test?format=xml 将返回 XML 数据。

② /test?format=json 将返回 JSON 数据。

根据参数来判断的缺点是，需要额外传递 format 参数，URL 变得冗余烦琐，缺少了 REST 的简洁风范。

使用内容协商的功能，如果不使用第三种方式，可以什么都不用配置，默认就能支持前面两种方式。下面介绍如果需要配置，应该如何配置。

① 需要在 Spring 的配置文件中进行如下配置。

```
<!-- Total customization - see below for explanation. -->
    <bean id="contentNegotiationManager" class="org.springframework.web.
    accept.ContentNegotiationManagerFactoryBean">
        <property name="favorPathExtension" value="false" />
        <property name="favorParameter" value="true" />
        <property name="parameterName" value="format" />
        <property name="ignoreAcceptHeader" value="false"/>
        <property name="useJaf" value="false"/>
        <property name="defaultContentType" value="text/html" />

        <property name="mediaTypes">
            <map>
                <entry key="json" value="application/json" />
                <entry key="xml" value="application/xml" />
            </map>
        </property>
    </bean>
```

① favorPathExtension 参数表示是否开启后缀，默认为 true（使用形如/ test /a.json、/ test /a.xml 的方式）。

② favorParameter 参数表示是否开启 request 参数识别，默认为 false（使用形如/ test /a?format=json、/ test /?format=xml 的方式）。

③ parameterName 参数表示使用参数的名字，默认为 format，如果配置为 mediaType，则请求格式变为/ test /a?mediaType=json。

④ ignoreAcceptHeader 参数表示是否关闭 accept 头识别，默认为 false，即默认开启 Accept 头识别。

⑤ defaultContentType 参数表示服务器默认的 mediaType 类型。

② 在 mvc:annotation-driven 里面使用内容协商。

```
<mvc:annotation-driven
content-negotiation-manager="contentNegotiationManager" />
```

③ 测试文件的变化。

- contentType : 'application/xml; charset=utf-8'，这句话可以注释掉。
- 在请求的 URL 上添加后缀和参数试试看。

如图 15-17 所示。

```
1 <%@ page language="java" contentType="text/html; charset=UTF-8" pageEncoding="UTF-8"%>
2 <script type="text/javascript" src="../js/jquery-2.1.1.min.js"></script>
3 <script>
4 $().ready(function(){
5     $.ajax({
6         url : '../hello6?format=xml',
7         type : 'post',
8         data : JSON.stringify({id:"1",username:"test"}),
9 //          contentType : 'application/xml; charset=utf-8',
10        error: function(){ alert('Error loading XMLdocument'); },
11        success: function(xml){
12            $(xml).find("testxml").children("pm").each(function(i){
13                var areaCode=$(this).children("areaCode").text();
14                var phoneNumber=$(this).children("phoneNumber").text();
15                alert("areaCode="+areaCode+",phoneNumber="+phoneNumber);
16            });
17        }
18    });
19 });
20 </script>
```

图 15-17　根据参数值判断数据类型的方式

本章小结

本章首先介绍了 Ajax 和 JSON 的概念，以及它们在 Spring MVC 中的应用场景。针对 Ajax 在 Spring MVC 中的应用分成了 3 种情况：直接的 Ajax 处理；通过注解进行 Ajax 处理；使用 ResponseEntity 和 HttpEntity 支持 Ajax。本章对这 3 种应用方式都做了详细介绍，同时针对另一种数据格式 XML 在 Spring MVC 中的应用也做了实际案例说明。本章还讲解了支持 JSON 和 XML 的幕后英雄 HttpMessageConverter，让读者对 Spring MVC 支持 Ajax 的基本原理有初步的了解。本章最后介绍了内容协商的 3 种方式。相信读者通过本章的学习，会对数据在 Spring MVC 中的传递格式有比较深的了解。

习题

1. 用于将HttpServletRequest的getInputStream内容绑定到入参的注解是____________。

2. 将Controller方法返回的对象通过适当的转换器转换为指定的格式，并写入response对象的body区的注解是____________。

上机指导

实现一个最直接的Ajax处理，从页面中提交用户id和用户名到Controller中，Controller通过model接收数据，在得到用户数据后将数据写入response，页面通过回调函数获取数据，显示数据。

16 第16章　文件的上传和下载

学习目标

- 了解 MultipartResolver
- 掌握 CommonsMultipartResolver 的实现方式
- 掌握 StandardServletMultipartResolver 的实现方式
- 掌握文件的上传方法
- 掌握文件的下载方法
- 掌握测试的方法

在互联网应用中，上传图片或者文件的操作是非常常见的，这些操作都涉及文件的上传和下载功能。Spring MVC 为文件的上传及下载提供了良好的支持。Spring MVC 的文件上传是通过 MultipartResolver（Multipart 解析器）处理的。

16.1　MultipartResolver 概述

MultipartResolver 概述

MultipartResolver 用于处理文件上传，当收到请求时，DispatcherServlet 的 checkMultipart 方法会调用 MultipartResolver 的 isMultipart 方法判断请求中是否包含文件。如果请求数据中包含文件，则调用 MultipartResolver 的 resolveMultipart 方法对请求的数据进行解析，然后将文件数据解析成 MultipartFile 并封装在 MultipartHttp ServletRequest（继承了 HttpServletRequest）对象中，最后传递给 Controller，在 MultipartResolver 接口中有如下方法。

- boolean isMultipart(HttpServletRequest request)；
- MultipartHttpServletRequest resolveMultipart(HttpServletRequest request)；
- void cleanupMultipart(MultipartHttpServletRequest request)。

MultipartFile 封装了请求数据中的文件，此时这个文件存储在内存中或临时的磁盘文件中，需要将其转存到一个合适的位置，因为请求结束后临时存储将被清空。在 MultipartFile 接口中有如下方法。

- String getName()，获取参数的名称；
- String getOriginalFilename()，获取文件的原名称；
- String getContentType()，获取文件内容的类型；
- boolean isEmpty()，文件是否为空；
- long getSize()，文件大小；
- byte[] getBytes()，将文件内容以字节数组的形式返回；
- InputStream getInputStream()，将文件内容以输入流的形式返回；
- void transferTo(File dest)，将文件内容传输到指定文件中。

MultipartResolver 是一个接口，它的实现类包括 CommonsMultipartResolver 类和 StandardServletMultipartResolver 类，如图 16-1 所示。

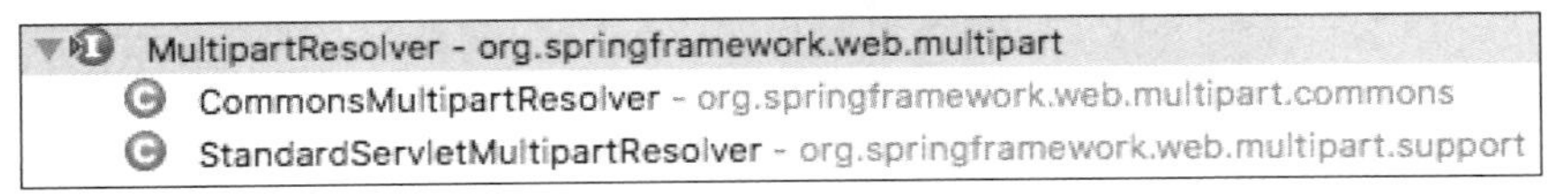

图 16-1　MultipartResolver 实现类

其中，CommonsMultipartResolver 使用 Commons FileUpload 来处理 multipart 请求，所以在使用时，必须引入相应的 JAR 包；而 StandardServletMultipartResolver 是基于 Servlet 3.0 来处理 multipart 请求的，所以不需要引用其他 JAR 包，但是必须使用支持 Servlet 3.0 的容器才可以。以 Tomcat 为例，Tomcat 从 7.0.x 的版本开始就支持 Servlet 3.0 了。

下面具体讲解两种实现文件上传及下载的方式。

16.2　CommonsMultipartResolver 实现方式

CommonsMultipartResolver 依赖于 Apache 下的 jakarta Commons FileUpload，可以在 Spring 的各个版本中使用，只是它要依赖于第三方的 JAR 包才能实现。下面介绍其具体使用方式。

CommonsMultipart-Resolver 实现方式

16.2.1　引入 JAR 包

引入以下 JAR 包：commons-fileupload-1.2.2.jar 和 commons-io-2.4.jar。

注意

JAR 包的版本不限定于如上所示。

16.2.2　配置文件

在 Spring MVC 配置文件中加入如下所示的配置。

```
    <!-- 定义文件上传解析器 -->
```

```
<bean id="multipartResolver" class="org.springframework.web.multipart. commons.
    CommonsMultipartResolver">
    <!-- 设定默认编码 -->
    <property name="defaultEncoding" value="UTF-8"></property>
    <!-- 设定文件上传的最大值为 5MB, 5*1024*1024 -->
    <property name="maxUploadSize" value="5242880"></property>
    <!-- 设定文件上传时写入内存的最大值，如果小于这个参数就不会生成临时文件，默认为
    10240 字节-->
    <property name="maxInMemorySize" value="40960"></property>
    <!-- 上传文件的临时路径 -->
    <property name="uploadTempDir" value="fileUpload/temp"></property>
    <!-- 延迟文件解析 -->
    <property name="resolveLazily" value="true"/>
</bean>
```

16.2.3 上传表单

多数文件上传都是通过表单形式提交给后台服务器的，因此，要实现文件上传功能，就需要提供一个文件上传的表单，而该表单必须满足以下 3 个条件。

（1）form 表单的 method 属性设置为 post。

（2）form 表单的 enctype 属性设置为 multipart/form-data。

（3）提供<input type="file" name="filename"/>的文件上传输入框。

当客户端 form 表单的 enctype 属性为 multipart/form-data 时，浏览器就会采用二进制流的方式来处理表单数据，服务器端就会对文件上传的请求进行解析处理。文件上传表单的示例代码如下。

```
<%@ page language="java" contentType="text/html; charset=UTF-8"
    pageEncoding="UTF-8"%>
<%
String path = request.getContextPath();
String basePath = request.getScheme()+"://"+request.getServerName()+":"+
    request.getServerPort()+path+"/";
%>
<!DOCTYPE meta PUBLIC "-//W3C//DTD HTML 4.01 Transitional//EN"
"http://www.w3.org/TR/html4/loose.dtd">
<html>
<head>
    <meta http-equiv="Content-Type" content="text/html; charset=UTF-8">
    <base href="<%=basePath%>">
</head>
<form action="test/file_upload" method="post" enctype="multipart/form-data">
    <input type="file" name="file">
    <input type="submit" value="submit">
    </form>
</body>
</html>
```

16.2.4 处理文件

处理文件的代码如下。

```
@RequestMapping("/file_upload")
    public String upload(@RequestParam(value = "file", required = false)
```

```
    MultipartFile file, HttpServletRequest request,Model mv) {
    // 文件不为空
    if(!file.isEmpty()) {
        String oldName=file.getOriginalFilename();
        //获取文件保存路径
        String realpath=request.getSession().getServletContext().getRealPath
        ("/fileUpload/");
        //设置文件路径
        File file2=new File(realpath);
        //检查路径是否存在
        if (!file2.isDirectory()) {
            file2.mkdirs();
        }
        //设置文件名
        String newName=UUID.randomUUID().toString()+oldName.substring
        (oldName.lastIndexOf("."), oldName.length());
    // 转存文件
    try {
    file.transferTo(new File(file2, newName));
    mv.addAttribute("fileNames", newName);
         return"success";
         } catch (IllegalStateException | IOException e) { e.printStackTrace();
         }
         }
    return"error";
    }
```

16.2.5 源码分析

CommonsMultipartResolver 实现了 MultipartResolver 接口，resolveMultipart 方法如下所示，其中 resolveLazily 判断是否要延迟解析文件（通过 XML 可以设置）。当 resolveLazily 为 false 时，会立即调用 parseRequest 方法对请求数据进行解析，然后将解析结果封装到 DefaultMultipart-HttpServletRequest 中。当 resolveLazily 为 true 时，会在 DefaultMultipartHttpServletRequest 的 initializeMultipart 方法调用 parseRequest 方法对请求数据进行解析，而 initializeMultipart 方法又被 getMultipartFiles 方法调用，即当需要获取文件信息时才会去解析请求数据，这种方式用了“懒加载”的思想。

```
@Override
public MultipartHttpServletRequest resolveMultipart(final HttpServletRequest
request) throws MultipartException {
    Assert.notNull(request, "Request must not be null");
    if (this.resolveLazily) {
        //懒加载，当调用 DefaultMultipartHttpServletRequest 的 getMultipartFiles
        方法时才解析请求数据
        return new DefaultMultipartHttpServletRequest(request) {
        //当 getMultipartFiles 方法被调用时，如果还未解析请求数据，则调用 initialize-
        Multipart 方法进行解析
        @Override
        protected void initializeMultipart() {
        MultipartParsingResult parsingResult = parseRequest(request);
```

```
            setMultipartFiles(parsingResult.getMultipartFiles());
            setMultipartParameters(parsingResult.getMultipartParameters());
            setMultipartParameterContentTypes(parsingResult.
            getMultipartParameterContentTypes());
        }
    };
    } else {
        //立即解析请求数据，并将解析结果封装到 DefaultMultipartHttpServletRequest 对
        象中
        MultipartParsingResult parsingResult = parseRequest(request);
        return new DefaultMultipartHttpServletRequest(request,
        parsingResult.getMultipartFiles(), parsingResult.getMultipart
        Parameters(), parsingResult.getMultipartParameterContentTypes());
        }
    }
```

从上面的代码中可以看到，对请求数据的解析工作是在 parseRequest 方法中进行的，继续看 parseRequest 方法源码。

```
protected MultipartParsingResult parseRequest(HttpServletRequest request)
    throws MultipartException {
    // 获取请求的编码类型
    String encoding = determineEncoding(request);
    FileUpload fileUpload = prepareFileUpload(encoding);
    try {
        List<FileItem> fileItems = ((ServletFileUpload)
            fileUpload).parseRequest(request);
        return parseFileItems(fileItems, encoding);
    } catch (...) {}
}
```

在 parseRequest 方法中，首先调用了 prepareFileUpload 方法根据编码类型确定一个 FileUpload 实例，然后利用这个 FileUpload 实例解析请求数据后得到文件信息，最后将文件信息解析成 CommonsMultipartFile（实现了 MultipartFile 接口）并包装在 MultipartParsingResult 对象中。

16.3 StandardServletMultipartResolver 实现方式

从 Spring 3.1 开始，Spring 就提供了 MultipartResolver 的实现 StandardServlet-MultipartResolver，用于处理 multipart 请求。下面以实例说明 StandardServlet-MultipartResolver 的用法。

StandardServlet-MultipartResolver 实现方式

16.3.1 配置文件

在 Spring 的核心配置文件中设置 Bean，Bean 的 id 必须是 MultipartResolver。这里并没有对上传文件的大小等参数进行配置，因为相应的配置都移到了 web.xml 中。要注意 XML 中 web-app 节点使用的版本，必须是 3.0 以上。

在 Spring 的核心配置文件中添加如下 Bean。

```
<bean id="multipartResolver" class="org.springframework.web.multipart.support.
StandardServletMultipartResolver"> </bean>
```

在 web.xml 文件中添加如下配置。

```
<?xml version="1.0" encoding="UTF-8"?>
<web-app version="3.0" xmlns="http://java.sun.com/xml/ns/javaee"
    xmlns:xsi="http://www.w3.org/2001/XMLSchema-instance"
    xsi:schemaLocation="http://java.sun.com/xml/ns/javaee
    http://java.sun.com/xml/ns/javaee/web-app_3_0.xsd">
    …
    <!--Spring MVC 前端控制器 DispatcherServlet -->
    <servlet>
        <servlet-name>springmvc_rest</servlet-name>

    <servlet-class>org.springframework.web.servlet.DispatcherServlet
    </servlet-class>
         <init-param>
              <param-name>contextConfigLocation</param-name>
              <param-value>classpath:/spring/springmvc.xml</param-value>
         </init-param>
         <load-on-startup>1</load-on-startup>
         <multipart-config>
    <!-- 临时文件目录 -->
    <location>d:/</location>
    <!-- 上传文件最大为 2MB -->
    <max-file-size>2097152</max-file-size>
    <!-- 上传文件整个请求不超过 4MB -->
    <max-request-size>4194304</max-request-size>
    </multipart-config>
    </servlet>
    …
</web-app>
```

16.3.2　上传表单

在<form>标签中加入 enctype="multipart/form-data"表示该表单要提交文件。具体代码如下所示。

```
<%@ page language="java" contentType="text/html; charset=UTF-8" pageEncoding=
    "UTF-8"%>
<%
String path = request.getContextPath();
String basePath = request.getScheme()+"://"+request.getServerName()+": "+
    request.getServerPort()+path+"/";
%>
<!DOCTYPE meta PUBLIC "-//W3C//DTD HTML 4.01 Transitional//EN" "http://www.
    w3.org/TR/html4/loose.dtd">
<html>
<head>
    <meta http-equiv="Content-Type" content="text/html; charset=UTF-8">
    <base href="<%=basePath%>">
</head>
<form action="test/file_upload" method="post" enctype="multipart/form-data">
```

```
    <input type="file" name="file">
    <input type="submit" value="submit">
    </form>
</body>
</html>
```

16.3.3 处理文件

Controller 处理文件可以用两种方式接收参数，如下所示。

1. MultipartFile 类型的参数

具体如下。

```
@RequestMapping("/file_upload")
public String  upload1(@RequestParam(value = "file", required = false)
MultipartFile file, HttpServletRequest request,Model mv) {
// 文件不为空
if(!file.isEmpty()) {
        String oldName=file.getOriginalFilename();
        //获取文件保存路径
        String realpath=request.getSession().getServletContext().getRealPath
        ("/fileUpload/");
        //设置文件路径
        File file2=new File(realpath);
        //检查路径是否存在
        if (!file2.isDirectory()) {
            file2.mkdirs();
        }
        //设置文件名
        String newName=UUID.randomUUID().toString()+oldName.
            substring(oldName.lastIndexOf("."), oldName.length());
// 转存文件
try {
file.transferTo(new File(file2, newName));
mv.addAttribute("fileNames", newName);
return"success";
      } catch (IllegalStateException | IOException e) {
      e.printStackTrace();
      }
      }
return"error";
}
```

2. MultipartHttpServletRequest 类型的参数

具体如下。

```
@RequestMapping("/file_upload")
public String upload(MultipartHttpServletRequest request,Model mv) {
MultipartFile file = request.getFile("file");
// 文件不为空
if(!file.isEmpty()) {
        String oldName=file.getOriginalFilename();
        //获取文件保存路径
        String realpath=request.getSession().getServletContext().getRealPath
        ("/fileUpload/");
```

```
        //设置文件路径
        File file2=new File(realpath);
        //检查路径是否存在
        if (!file2.isDirectory()) {
            file2.mkdirs();
        }
        //设置文件名
        String newName=UUID.randomUUID().toString()+oldName.
        substring(oldName.lastIndexOf("."), oldName.length());
    // 转存文件
    try {
    file.transferTo(new File(file2, newName));
    mv.addAttribute("fileNames", newName);
    return"success";
        } catch (IllegalStateException | IOException e) {
    e.printStackTrace();
        }
        }
    return"error";
    }
```

16.3.4 源码分析

StandardServletMultipartResolver 实现了 MultipartResolver 接口，resolveMultipart 方法如下所示，其中 resolveLazily 判断是否要延迟解析文件（通过 XML 可以设置）。

```
public MultipartHttpServletRequest resolveMultipart(HttpServletRequest request)
throws MultipartException {
    return new StandardMultipartHttpServletRequest(request, this.resolve Lazily);
}
public StandardMultipartHttpServletRequest(HttpServletRequest request, boolean
lazyParsing) throws MultipartException {
    super(request);
    // 判断是否立即解析
    if (!lazyParsing) {
        parseRequest(request);
    }
}
```

对请求数据的解析工作是在 parseRequest 方法中进行的，下面继续看 parseRequest 方法的源码。

```
private void parseRequest(HttpServletRequest request) {
    try {
        Collection<Part> parts = request.getParts();
        this.multipartParameterNames = new LinkedHashSet<String>(parts.size());
        MultiValueMap<String, MultipartFile> files = new
        LinkedMultiValueMap<String, MultipartFile>(parts.size());
        for (Part part : parts) {
            String disposition = part.getHeader(CONTENT_DISPOSITION);
            String filename = extractFilename(disposition);
            if (filename == null) {
                filename = extractFilenameWithCharset(disposition);
            }
            if (filename != null) {
```

```
                    files.add(part.getName(), new StandardMultipartFile(part,
            filename));
            } else {
                this.multipartParameterNames.add(part.getName());
            }
        }
        setMultipartFiles(files);
    } catch (Throwable ex) {}
}
```

parseRequest 方法利用了 servlet 3.0 的 request.getParts 方法获取上传文件，并将其封装到 MultipartFile 对象中。

16.4 上传多个文件

配置方式延用前面的配置方式即可，修改页面和后台处理部分，具体代码如下。

（1）上传表单

```
<%@ page language="java" contentType="text/html; charset=UTF-8" pageEncoding=
    "UTF-8"%>
<%
String path = request.getContextPath();
String basePath = request.getScheme()+"://"+request.getServerName()+
    ":"+request.getServer
    Port()+path+"/";
%>
<!DOCTYPE meta PUBLIC "-//W3C//DTD HTML 4.01 Transitional//EN"
"http://www.w3.org/TR/html4/loose.dtd">
<html>
<head>
    <meta http-equiv="Content-Type" content="text/html; charset=UTF-8">
    <base href="<%=basePath%>">
</head>
<form action="test/file_upload2" method="post" enctype="multipart/form-data">
    File 1:<input type="file" name="file">
    File 2:<input type="file" name="file">
    <input type="submit" value="submit">
    </form>
</body>
</html>
```

（2）处理文件

```
@RequestMapping(value = "/file_upload", method = RequestMethod.POST)
public String upload(
    HttpServletRequest request,
    @RequestParam(value = "file", required = false) MultipartFile[] files,
    Model mv) {
    String[] fileName = new String[files.length];
    try {
        for (int i = 0; i < files.length; i++) {
            if(files[i].getSize()>0){
                String a = FileUploadUtils.upload(request, files[i]);
```

```
                    String[] arr = a.split("/");
                    fileName[i] = arr[arr.length - 1];
                }
            }
        } catch (Exception e) {
            e.printStackTrace();
        }
        mv.addAttribute("fileNames", fileName);
        return "success";
    }
```

其中工具类 FileUploadUtils 的代码如下。

```
package com.inspur.ssm.util;

import java.io.File;
import java.io.IOException;
import java.io.UnsupportedEncodingException;

import javax.servlet.http.HttpServletRequest;

import org.springframework.web.multipart.MultipartFile;

public class FileUploadUtils {
        //默认大小为 50MB
        public static final long DEFAULT_MAX_SIZE = 52428800;
        //默认上传的地址
        public static String defaultBaseDir = "fileUpload";
        public static final String upload(HttpServletRequest
          request, MultipartFile file)throws Exception{
          String filename = extractFilename(file, defaultBaseDir);
          File desc = getAbsoluteFile(extractUploadDir(request), filename);
          file.transferTo(desc);
        return filename;
    }
    private static final File getAbsoluteFile(String uploadDir, String filename)
          throws IOException {
          if(uploadDir.endsWith("/")) {
              uploadDir = uploadDir.substring(0, uploadDir.length() - 1);
          }
          if(filename.startsWith("/")) {
              filename = filename.substring(0, uploadDir.length() - 1);
          }
          File desc = new File(uploadDir + "/" + filename);
    if(!desc.getParentFile().exists()) {
            desc.getParentFile().mkdirs();
        }
        if(!desc.exists()) {
            desc.createNewFile();
        }
         return desc;
      }
     public static final String extractFilename(MultipartFile file, String
          baseDir) throws UnsupportedEncodingException {
          String filename = file.getOriginalFilename();
          int slashIndex = filename.indexOf("/");
```

```
            // Check for Windows-style path
            int winSep = filename.lastIndexOf('\\');
            // Cut off at latest possible point
            int pos = (winSep > slashIndex ? winSep : slashIndex);
            if (pos != -1)  {
                // Any sort of path separator found...
                filename = filename.substring(pos + 1);
            }
            /*if (slashIndex >= 0) {
                 filename = filename.substring(slashIndex + 1);
            }*/
            filename = baseDir + "/" + filename;
            return filename;
        }
        public static final String extractUploadDir(HttpServletRequest request) {
            return request.getServletContext().getRealPath("/");
        }
    }
```

（3）成功显示页面

以上所有测试成功后返回的页面都是该页面，代码如下所示。

```
<%@ page language="java" contentType="text/html; charset=UTF-8"
pageEncoding="UTF-8"%>
<%@taglib prefix="c" uri="http://java.sun.com/jstl/core_rt"%>
<%
    String path = request.getContextPath();
    String basePath = request.getScheme()+"://"+request.getServerName()+":"+
    request.getServerPort()+path+"/";
%>
<!DOCTYPE HTML>
<html>
<head>
<base href="<%=basePath%>">
<meta http-equiv="Content-Type" content="text/html; charset=UTF-8">
<title></title>
</head>
<body>
    <c:forEach var="fileName" items="${fileNames}">
        <a href="test/download?fileName=${fileName}">${fileName}</a><br>
    </c:forEach>
</body>
</html>
```

16.5 文件下载

文件下载

我们也可以把上传的文件下载下来，下面介绍 Controller 的方法来实现。

```
@RequestMapping(value = "download", method = RequestMethod.GET)
    publicstaticvoid download(HttpServletRequest request,HttpServletResponse
        response, String fileName) throws Exception {
        response.setContentType("text/html;charset=UTF-8");
        request.setCharacterEncoding("UTF-8");
```

```
        BufferedInputStream bis = null;
        BufferedOutputStream bos = null;
        String newFileName = new String(fileName.getBytes("ISO8859-1"), "UTF-8");
        String ctxPath = request.getSession().getServletContext().
           getRealPath("/")+ FileUploadUtils.defaultBaseDir;
        String downLoadPath = ctxPath + "/" + newFileName;
        long fileLength = new File(downLoadPath).length();
        response.setHeader("Content-disposition", "attachment;filename="+ new
           String(newFileName.getBytes("UTF-8"), "ISO8859-1"));
        response.setHeader("Content-Length", String.valueOf(fileLength));
        bis = new BufferedInputStream(new FileInputStream(downLoadPath));
        bos = new BufferedOutputStream(response.getOutputStream());
        byte[] buff = newbyte[2048];
        int bytesRead;
        while (-1 != (bytesRead = bis.read(buff, 0, buff.length))) {
           bos.write(buff, 0, bytesRead);
        }
        bis.close();
        bos.close();
    }
```

16.6　测试

（1）upload.jsp 页面显示如图 16-2 所示。

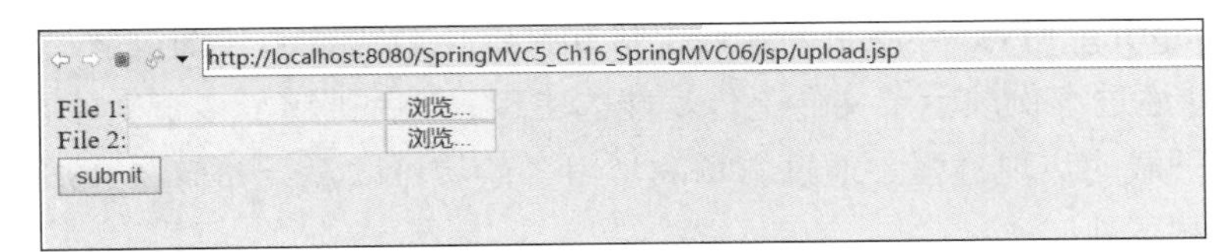

图 16-2　upload.jsp 页面显示

（2）单击“浏览”按钮，选择文件，如图 16-3 所示。

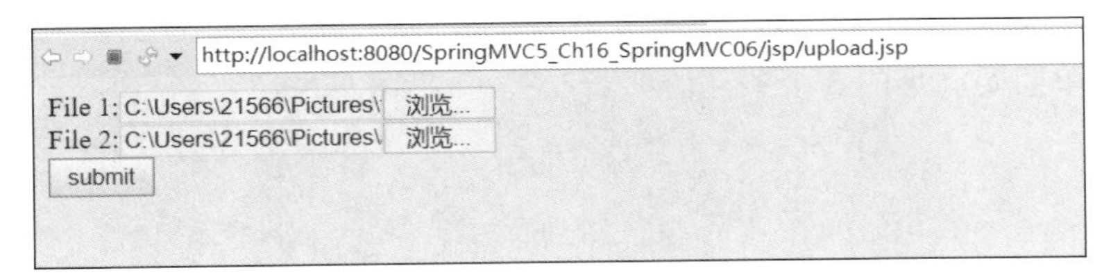

图 16-3　选择文件

（3）单击“submit”按钮后，上传文件成功界面如图 16-4 所示。

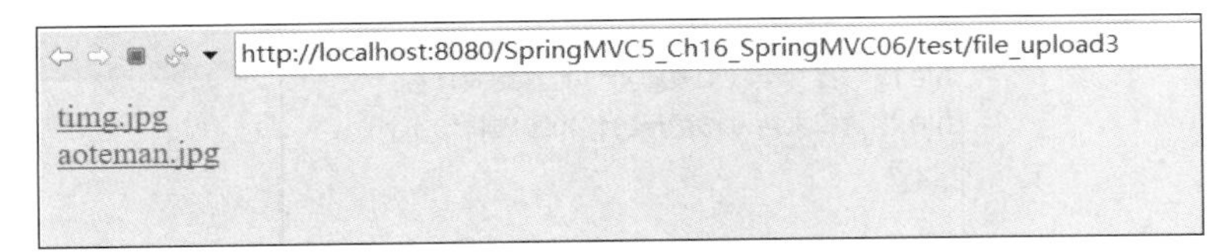

图 16-4　上传文件成功页面

（4）单击文件的链接，就可以下载文件了，如图 16-5 所示。

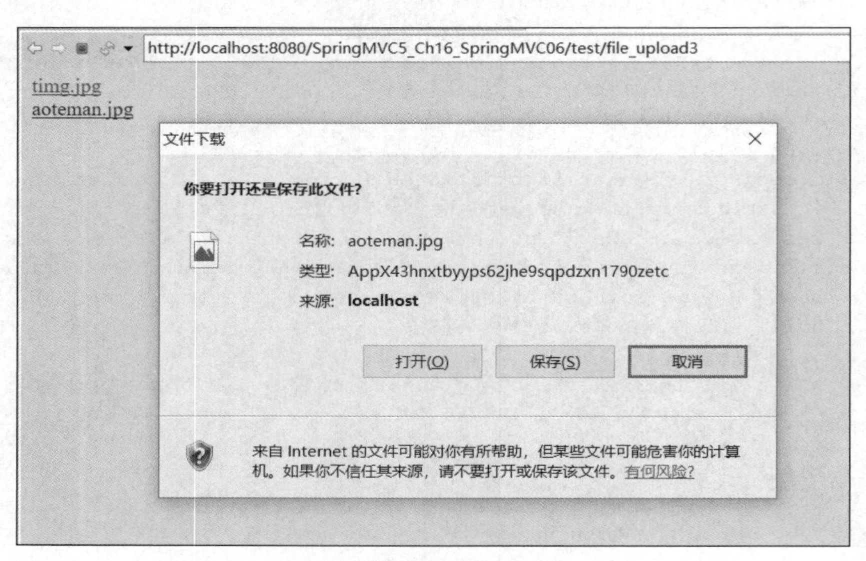

图 16-5　文件下载页面

本章小结

本章主要对 Spring MVC 环境下的文件上传和文件下载进行了详细讲解。首先讲解了如何实现文件上传，并通过案例演示了文件上传功能的实现；然后讲解了多个文件上传的实现过程；最后讲解了文件下载的实现过程。通过本章的学习，读者可以学会如何在 Spring MVC 环境下进行文件的上传和下载。

习题

简述上传表单需要满足的 3 个条件。

上机指导

书写一个项目，实现文件上传和下载的功能，示例如下。

（1）选择待上传的文件（见图 16-6）。

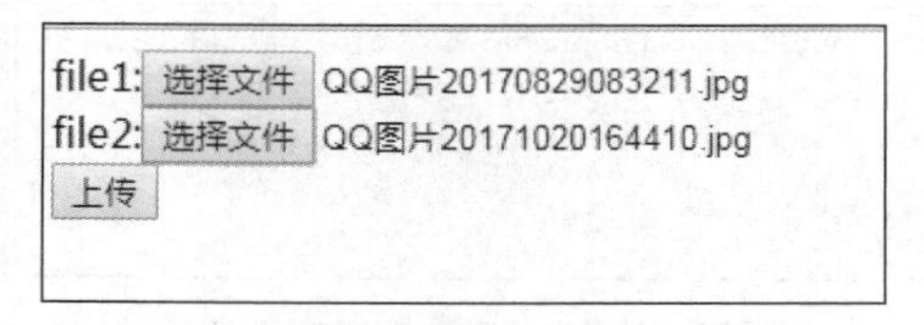

图 16-6　选择上传文件

（2）上传成功后显示图片名称，然后生成超链接（见图 16-7）。

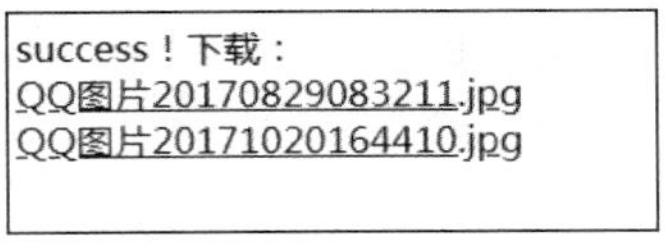

success！下载：
QQ图片20170829083211.jpg
QQ图片20171020164410.jpg

图 16-7　上传成功

（3）单击超链接后可以下载该图片（见图 16-8）。

QQ图片2017082....jpg
QQ图片2017102....jpg

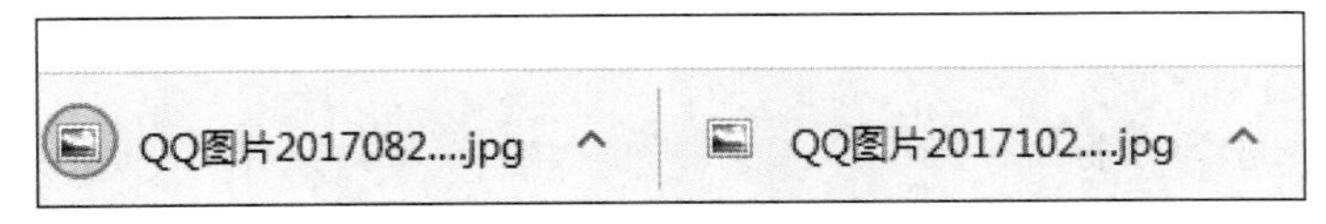

图 16-8　下载图片

第17章　SSM框架整合

学习目标

- 理解三大框架的基本概念
- 理解三大框架的整合思路
- 掌握环境准备的方法
- 了解三大框架的工程结构
- 掌握三大框架的整合过程及方法

通过前面的讲解，相信读者已经掌握了 Spring、Spring MVC 和 MyBatis 框架的使用方法，并且理解了各个框架的基本理论知识。在前面已经介绍过 Spring 和 MyBatis 的整合，但是在实际项目开发中，这三大框架通常都是整合在一起使用的，所以本章就针对 SSM（Spring、Spring MVC 和 MyBatis）三大框架的整合使用进行详细解说。

17.1　三大框架的基本概念

为了更好地进行三个框架的融合，我们先来回顾三个框架的基本概念，理解每个框架所担任的角色和承担的主要功能。

三大框架的基本概念

Spring 是一个开源框架，是于 2003 年兴起的一个轻量级的 Java 开发框架。它是为了解决企业应用开发的复杂性而创建的。

Spring MVC 属于 SpringFrameWork 的后续产品，已经融合在 Spring Web Flow 里面。Spring MVC 分离了控制器、模型对象、分派器以及处理程序对象的角色，这种分离让它们更容易进行定制。

MyBatis 是一个基于 Java 的持久层框架，它的前身是 iBatis。MyBatis 继承了 iBatis 的优秀功能。iBatis 提供的持久层框架包括 SQL Maps 和 DAO。这也使得 iBatis 消除了绝大多数的 JDBC 代码、参数的手工设置和结果集的检索，使数据持久化变得非常简单和方便。

17.2　整合思路

由于 Spring MVC 是 Spring 框架中的一个模块，所以 Spring MVC 和 Spring 之间不存在整合的问题，只要引入相关的 JAR 包就可以直接使用。因此 SSM 框架的整合主要是涉及 Spring 和 MyBatis 之间的整合，以及 Spring MVC 和 MyBatis 之间的整合。三大框架的整合思路如图 17-1 所示。

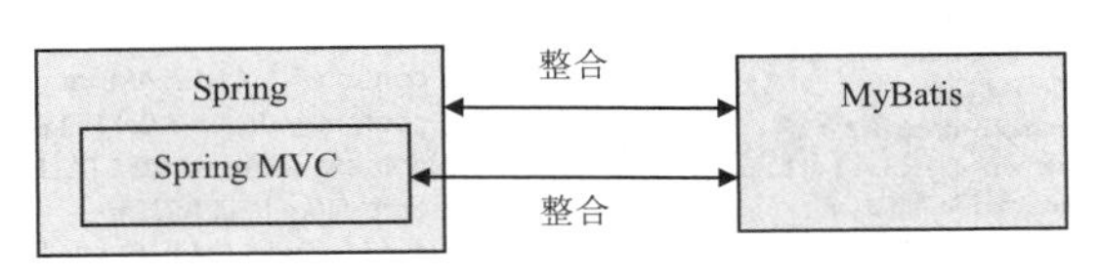

图 17-1　三大框架整合思路

SSM 系统架构逻辑如下：Spring 对各层进行整合，通过 Spring 管理数据持久层的 mapper（相当于 DAO 接口）；通过 Spring 管理业务逻辑层 service，service 中可以调用 mapper 接口；通过 Spring 进行事务控制；通过 Spring 管理表现层 Handler，Handler 中可以调用 service 接口。其中的 mapper、service、Handler 都是 JavaBean。整合后的数据执行流程图如图 17-2 所示。

（1）第一步：整合 DAO 层。

① MyBatis 和 Spring 进行整合，通过 Spring 管理 mapper 接口。

② 使用 mapper 扫描器自动扫描 mapper 接口并在 Spring 中进行注册。

（2）第二步：整合 service 层。

① 通过 Spring 管理 service 接口。

② 使用配置的方式将 service 接口配置在 Spring 配置文件中。

③ 实现事务控制。

（3）第三步：整合 Spring MVC。

由于 Spring MVC 是 Spring 的模块，不需要和 Spring 整合，直接把 Spring MVC 加入项目中即可。

下面通过 SSM 框架实现一个登录样例，以帮助读者更好地理解框架整合。

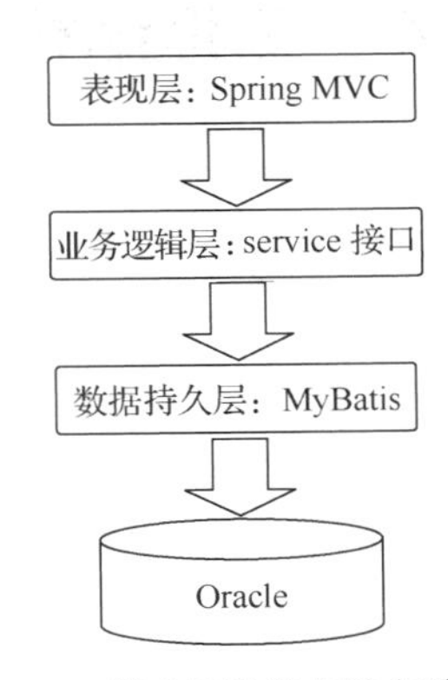

图 17-2　整合后的数据执行流程图

17.3　环境准备

本书所准备的环境如下。

（1）数据库环境：Oracle 12c，准备一个用户表，如“userssm”有用户名和用户密码字段，表里插入几条用户数据，供登录验证使用。

（2）Java 环境：JDK（Java 开发工具包）提供的 Java 开发环境和运行环境，是所有 Java 应用程序的基础。JDK 包含一组 API 和 JRE，这些 API 是构建 Java 应用程序的基础，而 JRE 是运行 Java 应用程序的基础。JDK 的版本需要在 8.0 以上，本书使用 JDK 8 版本进行讲解。

（3）Spring MVC 版本：Spring 5 及以上版本。

（4）所需要的 JAR 包：要实现 SSM 框架的整合，首先要准备这三个框架的 JAR 包，以及其他整合所需的 JAR 包。我们在前面讲解 Spring 与 MyBatis 框架整合时，已经介绍了 Spring 与 MyBatis 整合所需要的 JAR 包，这里只需要再加入 Spring MVC 的相关 JAR 包即可，具体的 SSM 整合需要的全部 JAR 包如图 17-3 所示。

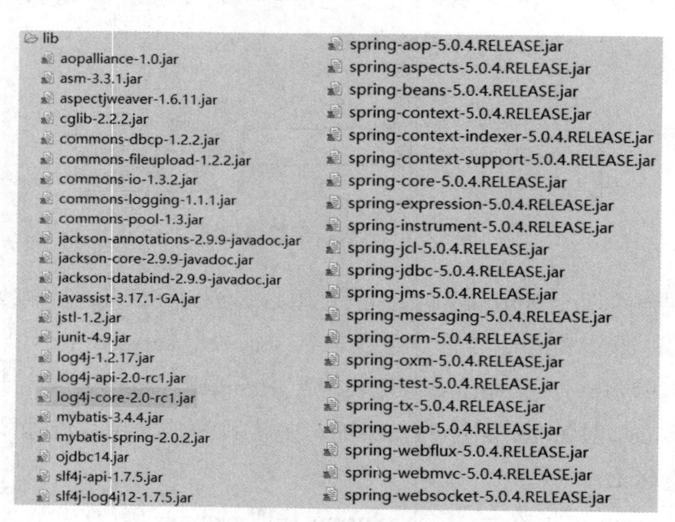

图 17-3　SSM 整合需要的 JAR 包

17.4　工程结构

本案例的工程结构如图 17-4 所示。

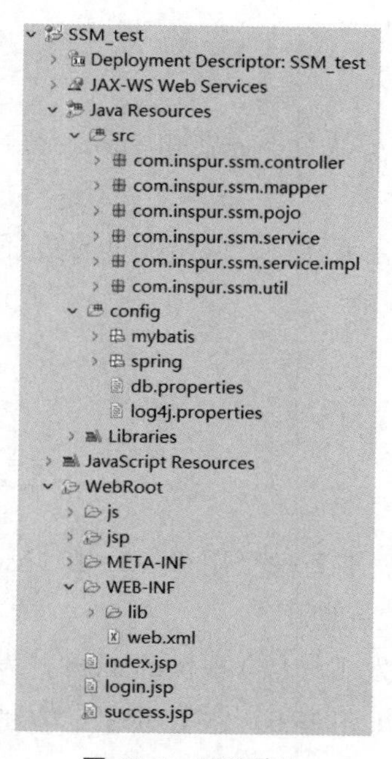

图 17-4　工程结构

17.5 三大框架的整合过程

三大框架的
整合过程

17.5.1 整合 MyBatis 和 Spring

在 Eclipse 中，创建一个名为 SSM_test 的动态 Web 项目，将整合所需要的 JAR 包添加到 lib 目录中。

在 SSM_test 项目下，创建一个名为 config 的源文件夹，在该文件夹中分别创建数据库常量配置文件 db.properties、Spring 配置文件 applicationContext.xml 以及 MyBatis 的配置文件 sqlMapConfig.xml，部分配置文件的代码如下所示。

1. sqlMapConfig.xml

它是 MyBatis 的配置文件，在该文件中只需要根据 POJO 类路径进行别名配置即可，数据源信息以及 mapper 接口文件扫描器将在 Spring 配置文件中配置。

```
<?xml version="1.0" encoding="UTF-8"?>
<!DOCTYPE configuration
PUBLIC "-//mybatis.org//DTD Config 3.0//EN"
"http://mybatis.org/dtd/mybatis-3-config.dtd">
<configuration>
    <!-- 配置别名 -->
    <typeAliases>
        <!-- 批量扫描别名 -->
        <package name="com.inspur.ssm.pojo"></package>
    </typeAliases>
</configuration>
```

2. applicationContext-dao.xml

该配置文件中主要包含以下 3 个部分的内容。

（1）数据源。读取 db.properties 文件的配置和数据源配置，然后配置事务管理器并开启了事务注解。

（2）SqlSessionFactory。配置用于整合 MyBatis 框架的 MyBatis 工厂信息。

（3）mapper 扫描器。用来扫描 DAO 层以及 Service 层的配置。

具体代码如下所示。

```
<?xml version="1.0" encoding="UTF-8"?>
<beans xmlns="http://www.springframework.org/schema/beans"
xmlns:xsi="http://www.w3.org/2001/XMLSchema-instance"
xmlns:p="http://www.springframework.org/schema/p"
xmlns:mvc="http://www.springframework.org/schema/mvc"
xmlns:context="http://www.springframework.org/schema/context"
xmlns:aop="http://www.springframework.org/schema/aop"
xmlns:tx="http://www.springframework.org/schema/tx"
xsi:schemaLocation="http://www.springframework.org/schema/beans
http://www.springframework.org/schema/beans/spring-beans.xsd
http://www.springframework.org/schema/mvc
http://www.springframework.org/schema/mvc/spring-mvc.xsd
http://www.springframework.org/schema/context
http://www.springframework.org/schema/context/spring-context.xsd
```

```
    http://www.springframework.org/schema/aop
    http://www.springframework.org/schema/aop/spring-aop.xsd
    http://www.springframework.org/schema/tx
    http://www.springframework.org/schema/tx/spring-tx.xsd">
        <!-- 加载配置文件 db.properties 中的内容，db.properties 文件中 key 命名要有一定命名
            规则 -->
        <context:property-placeholder location="classpath:db.properties" />
        <!-- 数据源，使用 dbcp -->
        <bean id="dataSource" class="org.apache.commons.dbcp.BasicDataSource"
            destroy-method="close">
            <property name="driverClassName" value="${jdbc.driver}" />
            <property name="url" value="${jdbc.url}" />
            <property name="username" value="${jdbc.username}" />
            <property name="password" value="${jdbc.password}" />
            <property name="maxActive" value="10" />
            <property name="maxIdle" value="5" />
        </bean>

        <!-- sqlSessionfactory -->
        <bean id="sqlSessionFactory" class="org.mybatis.spring.SqlSessionFactoryBean">
            <!-- 加载 MyBatis 的配置文件 -->
            <property  name="configLocation" value="classpath:mybatis/sqlMapConfig
            .xml"></property>
            <!-- 数据源 -->
            <property name="dataSource" ref="dataSource"></property>
        </bean>

        <!-- mapper 批量扫描 -->
        <bean class="org.mybatis.spring.mapper.MapperScannerConfigurer">
            <!-- 扫描包路径，如果需要扫描多个包，中间使用半角逗号隔开 -->
            <property name="basePackage" value="com.inspur.ssm.mapper"></property>
            <property name="sqlSessionFactoryBeanName" value="
            sqlSessionFactory"></property>
        </bean>
</beans>
```

3. pojo 类

具体代码如下。

```
package com.inspur.ssm.pojo;
public class Userssm {
    private String id;
    private String username;
    private String password;
    public String getId() {
        return id;
    }
    public void setId(String id) {
        this.id = id;
    }
    public String getUsername() {
        return username;
    }
    public void setUsername(String username) {
```

```
            this.username = username;
        }
        public String getPassword() {
            return password;
        }
        public void setPassword(String password) {
            this.password = password;
        }
    }
```

4. Mapper.java

具体代码如下。

```
package com.inspur.ssm.mapper;

import com.inspur.ssm.pojo.Userssm;

public interface UserssmMapper {
    UserssmfindUser(UserssmuserssmQuery);
}
```

5. Mapper.xml

具体代码如下。

```
<?xml version="1.0" encoding="UTF-8" ?>
<!DOCTYPE mapper PUBLIC "-//mybatis.org//DTD Mapper 3.0//EN" "http://
mybatis.org/dtd/mybatis-3-mapper.dtd" >
<mapper namespace="com.inspur.ssm.mapper.UserssmMapper" >
<select id="findUser" resultType="userssm" parameterType="userssm" >
    select  t1.id,
    t1.username,
    t1.role,
    t1.realname,
    t1.password
    from  USERSSM t1
    where username = #{username}
    and password = #{password}
</select>
</mapper>
```

17.5.2 Spring 整合 service

让 Spring 管理 service 接口。

1. 定义 service 接口

具体代码如下。

```
package com.inspur.ssm.service;

import javax.annotation.Resource;

import com.inspur.ssm.pojo.Userssm;

@Resource
public interface LoginService {
    //用户查询列表
```

```
    public UserssmfindUser(Userssmuserssm)throws Exception;

}
```

2. 定义 service 实现类

具体代码如下。

```
package com.inspur.ssm.service.impl;

import org.springframework.beans.factory.annotation.Autowired;
import org.springframework.stereotype.Service;

import com.inspur.ssm.mapper.UserssmMapper;
import com.inspur.ssm.pojo.Userssm;
import com.inspur.ssm.service.LoginService;

@Service("loginService")
public class LoginServiceImpl implements LoginService {
    @Autowired
    private UserssmMapperuserssmMapper;
    @Override
    public UserssmfindUser(UserssmuserssmQuery) throws Exception {
        Userssmuserssm = userssmMapper.findUser(userssmQuery);
        if(userssm == null){
            throw new Exception("用户信息不存在");
        }else{
            return userssm;
        }
    }

}
```

3. 在 Spring 容器中配置 service(applicationContext-service.xml)

```
<!- 原来 service 配置方式 -->
<bean id="loginService" class="com.inspur.ssm.service.impl.LoginServiceImpl">
</bean>
```

后续整合 Spring MVC，可以使用包扫描器将 service 配置取代（该步骤可以省略）。

4. 事务控制（applicationContext-transaction.xml）

具体代码如下。

```
<?xml version="1.0" encoding="UTF-8"?>
<beans xmlns="http://www.springframework.org/schema/beans"
    xmlns:xsi="http://www.w3.org/2001/XMLSchema-instance"
    xmlns:p="http://www.springframework.org/schema/p"
    xmlns:mvc="http://www.springframework.org/schema/mvc"
    xmlns:context="http://www.springframework.org/schema/context"
    xmlns:aop="http://www.springframework.org/schema/aop"
    xmlns:tx="http://www.springframework.org/schema/tx"
    xsi:schemaLocation="http://www.springframework.org/schema/beans
    http://www.springframework.org/schema/beans/spring-beans.xsd
    http://www.springframework.org/schema/mvc
    http://www.springframework.org/schema/mvc/spring-mvc.xsd
    http://www.springframework.org/schema/context
```

```
    http://www.springframework.org/schema/context/spring-context.xsd
    http://www.springframework.org/schema/aop
    http://www.springframework.org/schema/aop/spring-aop.xsd
    http://www.springframework.org/schema/tx
    http://www.springframework.org/schema/tx/spring-tx.xsd">

    <!-- 事务管理器 对MyBatis操作数据库的事务控制，Spring使用JDBC的事务控制类 -->
    <bean id="transactionManager"
        class="org.springframework.jdbc.datasource.DataSourceTransaction
        Manager">
        <!-- 数据源 DataSource在applicationContext-dao.xml中配置了 -->
        <property name="dataSource" ref="dataSource"></property>
    </bean>
    <!-- 通知 -->
    <tx:advice id="txAdvice" transaction-manager="transactionManager">
        <tx:attributes>
            <tx:method name="save*" propagation="REQUIRED" />
            <tx:method name="delete*" propagation="REQUIRED" />
            <tx:method name="insert*" propagation="REQUIRED" />
            <tx:method name="update*" propagation="REQUIRED" />
            <tx:method name="find*" propagation="SUPPORTS" read-only="true" />
            <tx:method name="get*" propagation="SUPPORTS" read-only="true" />
            <tx:method name="select*" propagation="SUPPORTS" read-only="true" />
        </tx:attributes>
    </tx:advice>
    <!-- aop -->
    <aop:config>
        <aop:advisor advice-ref="txAdvice" pointcut="execution(*
          com.inspur.ssm.service.impl.*.*(..))" />
    </aop:config>

</beans>
```

17.5.3 整合Spring MVC

1. Spring MVC配置

创建springmvc.xml文件，配置处理器、映射器、适配器、视图解析器。

```
<?xml version="1.0" encoding="UTF-8"?>
<beans xmlns="http://www.springframework.org/schema/beans"
    xmlns:xsi="http://www.w3.org/2001/XMLSchema-instance"
    xmlns:p="http://www.springframework.org/schema/p"
    xmlns:mvc="http://www.springframework.org/schema/mvc"
    xmlns:context="http://www.springframework.org/schema/context"
    xmlns:aop="http://www.springframework.org/schema/aop"
    xmlns:tx="http://www.springframework.org/schema/tx"
    xsi:schemaLocation="http://www.springframework.org/schema/beans
    http://www.springframework.org/schema/beans/spring-beans.xsd
    http://www.springframework.org/schema/mvc
    http://www.springframework.org/schema/mvc/spring-mvc.xsd
    http://www.springframework.org/schema/context
    http://www.springframework.org/schema/context/spring-context.xsd
    http://www.springframework.org/schema/aop
    http://www.springframework.org/schema/aop/spring-aop.xsd
```

```
    http://www.springframework.org/schema/tx
    http://www.springframework.org/schema/tx/spring-tx.xsd">
    <!-- 可以扫描 Controller、service、DAO 等 Java 对象 -->
    <context:component-scan base-package="com.inspur.ssm.*"></context:
    component-scan>

      <!-- 静态资源解析 包括：js、css、img、html... -->
      <mvc:resources location="/js/" mapping="/js/**"></mvc:resources>
      <mvc:resources location="/css/" mapping="/css/**"></mvc:resources>
      <mvc:resources location="/images/" mapping="/images/**"></mvc:resources>
      <mvc:resources location="/html/" mapping="/html/**"></mvc:resources>

      <!-- 使用 mvc:annotation-driven 代替注解映射器和注解适配器配置 -->
      <mvc:annotation-driven />

      <!-- 视图解析器：解析 JSP，默认使用 JSTL 标签库，classpath 下有 JSTL 的包 -->
      <bean class="org.springframework.web.servlet.view.InternalResourceView
      Resolver">
          <property name="viewClass" value="org.springframework.web.servlet.
          view.JstlView" />
          <property name="prefix" value="/" />
          <property name="suffix" value=".jsp" />
      </bean>
      <!-- 配置处理文件上传的 Bean -->
      <bean id="multipartResolver" class="org.springframework.web.
      multipart.commons.CommonsMultipartResolver">
      <property name="maxUploadSize" value="104857600"/>
      </bean>
</beans>
```

2. 编写 Controller（即 Handler）

先使用 Spring 的注解@Controller 来标识控制器类，然后通过@Autowired 注解将 LoginService 接口对象注入本类中，最后编写一个根据用户名称和密码查询用户详情的方法 login，该方法会将获取的客户详情返回到视图名为 success 的 JSP 页面中。

具体代码如下所示。

```
package com.inspur.ssm.controller;

import javax.servlet.http.HttpServletRequest;
import javax.servlet.http.HttpSession;

import org.springframework.beans.factory.annotation.Autowired;
import org.springframework.stereotype.Controller;
import org.springframework.web.bind.annotation.RequestMapping;

import com.inspur.ssm.pojo.Userssm;
import com.inspur.ssm.service.LoginService;

@Controller
public class LoginController {
@Autowired
private LoginServiceloginService;
```

```
    // 登录
    @RequestMapping("/login")
    public String login(HttpServletRequest request, Userssmuserssm)
        throws Exception {
        Userssm user = null;

        try {
            // 调用service进行用户身份验证
            user = loginService.findUser(userssm);
        } catch (Exception e) {
            e.printStackTrace();
            request.setAttribute("message", e.getMessage() );
        }
        if (user != null) {
            HttpSession session = request.getSession();
            // 在session中保存用户身份信息
            session.setAttribute("user", user);
            // 重定向到登录成功页面
            return "success";
        }
        return "login";
    }
}
```

17.5.4 配置前端控制器

1. Spring 配置

具体代码如下。

```
<!-- 加载Spring容器 -->
<context-param>
   <param-name>contextConfigLocation</param-name>
   <param-value>classpath:/spring/applicationContext-*.xml</param-value>
</context-param>
<listener>
   <listener-class>org.springframework.web.context.ContextLoaderListener
</listener-class>
</listener>
```

2. Spring MVC 配置

具体代码如下。

```
<!-- Spring MVC前端控制器DispatcherServlet -->
<servlet>
    <servlet-name>springmvc_rest</servlet-name>
    <servlet-class>org.springframework.web.servlet.DispatcherServlet
    </servlet-class>
    <init-param>
         <param-name>contextConfigLocation</param-name>
         <param-value>classpath:/spring/springmvc.xml</param-value>
    </init-param>
    <load-on-startup>1</load-on-startup>
</servlet>
<servlet-mapping>
```

```
        <servlet-name>springmvc_rest</servlet-name>
        <url-pattern>/</url-pattern>
    </servlet-mapping>
```

3. Spring 提供的解决乱码的方式

具体代码如下。

```
    <filter>
         <filter-name>CharacterEncodingFilter</filter-name>
         <filter-class>org.springframework.web.filter.CharacterEncodingFilter
</filter-class>
    <init-param>
    <param-name>encoding</param-name>
    <param-value>UTF-8</param-value>
    </init-param>
    </filter>
    <filter-mapping>
    <filter-name>CharacterEncodingFilter</filter-name>
    <url-pattern>/*</url-pattern>
    </filter-mapping>
```

17.5.5 编写页面

登录页面 login.jsp 的代码如下，用于展示用户登录输入框。

```
    <%@ page language="java" import="java.util.*" pageEncoding="UTF-8"%>
    <%
    String path = request.getContextPath();
       String basePath = request.getScheme()+"://"+request.getServerName()+":"+
       request.getServerPort()+path+"/";
    %>
    <!DOCTYPE HTML PUBLIC "-//W3C//DTD HTML 4.01 Transitional//EN">
    <html>
    <head>
    <base href="<%=basePath%>">
    <title>My JSP 'login.jsp' starting page</title>
    <meta http-equiv="pragma" content="no-cache">
    <meta http-equiv="cache-control" content="no-cache">
    <meta http-equiv="expires" content="0">
    <meta http-equiv="keywords" content="keyword1,keyword2,keyword3">
    <meta http-equiv="description" content="This is my page">
    </head>

    <body>
       <div>
           <div>
               <span style="color: red">${message}</span>
           </div>
           <form action="<%=path %>/login.action" method="post" name="ThisForm">
               <table>
                   <tr>
                       <td>账 户：</td>
                       <td><input class="username" name="username" type="text" /></td>
                   </tr>
                   <tr>
                       <td>密 码：</td>
```

```
                <td><input class="password" name="password" type="
            password"/></td>
            </tr>
            <tr>
                <td></td>
                <td><input type="submit" class="login-btn" value="登录"
                /></td>
            </tr>
        </table>
    </form>
  </div>
</body>
</html>
```

登录成功页面 success.jsp 的代码如下所示，通过 EL 表达式获取后台控制层返回的客户信息。

```
<%@ page language="java" import="java.util.*" pageEncoding="UTF-8"%>
<!DOCTYPE HTML PUBLIC "-//W3C//DTD HTML 4.01 Transitional//EN">
<html>
<head>
<title>My JSP 'success.jsp' starting page</title>
</head>
<body>
    This is my success page. <br>
    id:${user.id}<br>
    usersname:${user.username}<br>
    password:${user.password}<br>
</body>
</html>
```

17.5.6 数据库配置和日志配置

数据库配置文件 db.properties 示范：

```
jdbc.driver=oracle.jdbc.driver.OracleDriver
jdbc.url=jdbc:oracle:thin:@localhost:1521:xe
jdbc.username=zhou
jdbc.password=123456
```

日志配置文件 log4j.properties 示范：

```
log4j.appender.stdout=org.apache.log4j.ConsoleAppender
log4j.appender.stdout.Target=System.out
log4j.appender.stdout.layout=org.apache.log4j.PatternLayout
log4j.appender.stdout.layout.ConversionPattern=%d{ABSOLUTE} %5p [%t] %c{1}:
    %L - %m%n
#在开发环境下日志级别要设置成 DEBUG，生产环境设置成 info 或 error
log4j.rootLogger=DEBUG, stdout
```

17.5.7 项目部署

完成以上 SSM 框架整合的环境搭建工作以后，将动态项目发布到 Tomcat 服务器，启动服务器，在浏览器中输入访问地址 http://localhost:8080/SSM_test/login.jsp 后，其显示效果如图 17-5 所示。

访问成功页面如图 17-6 所示，已成功查询出 userssm 表中用户名为 inspur 的用户信息，这就说明 SSM 框架整合成功。

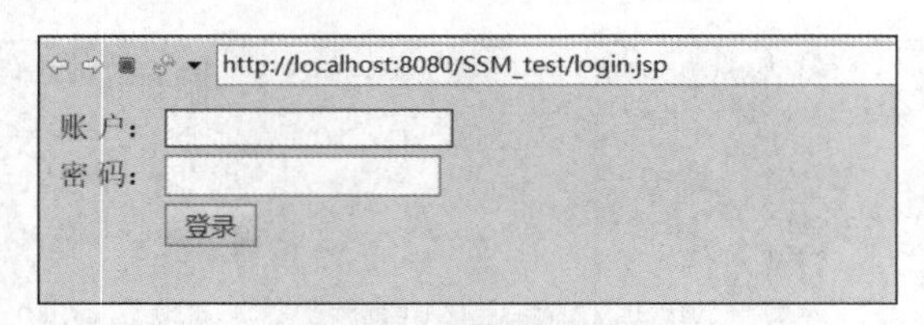

图 17-5　登录页面显示效果

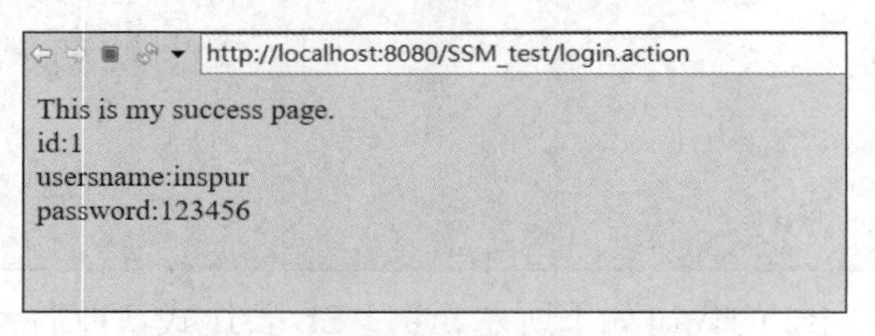

图 17-6　访问成功页面显示效果

本章小结

通过本章的学习，读者能够了解 SSM 框架的整合思路、掌握 SSM 框架的整合过程以及如何使用 SSM 框架。框架的整合是 SSM 框架使用的基础知识，读者一定要熟练掌握 SSM 框架的整合过程。

习题

1. SSM 框架中的 SSM 分别指的是______、______、______。
2. 下面说法错误的是（　　）。
   A. Spring 是一个轻量级的控制反转（IoC）和面向切面编程（AOP）的容器框架
   B. Spring MVC 分离了控制器、模型对象、分派器以及处理程序对象的角色，这种分离让它们更容易进行定制
   C. MyBatis 能将 POJO 对象映射成数据库中的记录
   D. 工程项目中 service 层的一般内容是接收页面请求的
3. 简述 SSM 框架整合的思路。
4. 简述 SSM 框架整合时，Spring 配置文件中的配置信息（无须写代码，只需要简单表述所要配置的内容即可）。

上机指导

在数据库中创建一个商品表 items(id,name,price)，并插入几条商品信息。使用 SSM 框架编写一个查询案例，根据商品 id 查询商品信息并将信息显示在页面上，如果商品 id 不存在，则显示提示信息“该商品信息不存在”。

18 第18章　医疗信息系统

学习目标

- 了解项目背景及项目结构
- 了解系统架构和文件组织结构
- 熟悉系统环境搭建的步骤
- 掌握用户登录模块和用户管理模块功能代码的编写方法

本章将通过前面介绍的 SSM（Spring+Spring MVC+MyBatis）框架知识来实现一个简易的医疗信息系统。在本章内容中，由于篇幅限制仅实现了用户登录模块和用户管理模块，读者学习完本章内容后，应该具有根据这两个模块完成其他模块的能力。该系统在开发过程中，整合了三大框架，并且在整合的基础上实现了系统的用户登录模块以及用户管理模块，对前面所学是很好的总结和回顾，同时使读者能熟练地将 SSM 相关技术应用于实际开发中。

医疗信息系统

18.1　项目背景及项目结构

本章将以一个医疗信息系统为例，为读者示范如何开发 Java EE 应用，该系统包含用户管理、科室管理、医生管理、挂号管理、药品发放等模块。

18.1.1　项目背景

医疗信息系统（Hospital Information System，HIS）在国际学术界已公认为新兴的医疗信息学的重要分支。美国该领域的著名学者莫里斯·科伦（Morris Collen）于 1988 年曾为“医疗信息系统”做了如下定义：利用电子计算机和通信设备，为医疗所属部门提供病人诊疗信息和行政管理信息的收集、存储、处理、提取和数据交换的能力，并满足所有授权用户的功能需求。

我国对“医疗信息系统”的定义：医疗信息系统是指利用计算机软硬件技术、网络通信技术等现代化手段，对医疗及其所属各部门在人流、物流、财流方面进行综合管理，对在医疗活动各阶段中产生的数据进行采集、存储、处理、提取、传输、汇总、加工生成各种信息，从而为医疗的整体运行提供全面的、自动化的管理及各种服务的信息系统。

18.1.2　程序框架结构图

系统采用了 SSM 架构，我们可将之划分为表现层、业务逻辑层（service 层）、数据持久层（DAO 层）。

（1）表现层：该层主要包括 Spring MVC 中的 Controller 类和 JSP 页面。Controller 类主要负责拦截用户请求，并调用业务逻辑层中相应组件的业务逻辑方法来处理用户请求，然后将相应的结果返回给 JSP 页面。

（2）业务逻辑层：该层由若干 service 接口和实现类组成。在本系统中，业务逻辑层接口的命名方式统一使用 service 结尾，其实现类的名称统一在接口名后加 Impl。该层主要用于实现系统的业务逻辑。

（3）数据持久层：该层由若干 DAO 接口和 MyBatis 映射文件组成。接口的名称统一以 DAO 结尾，且 MyBatis 的映射文件名称要与接口的名称相同。

各层次之间的关系如图 18-1 所示。

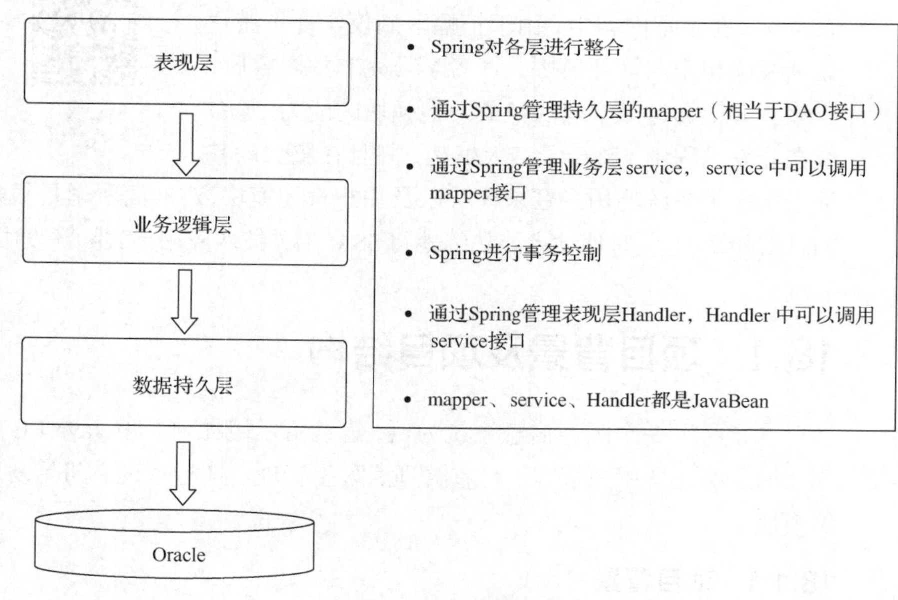

图 18-1　程序框架结构图

18.1.3　系统模块结构图

系统模块结构图如图 18-2 所示。

18.1.4　数据库的设计

使用 Oracle 建立一个数据库 orcl，该数据库共有 13 个表，以下是这些表的名称、结构和用途。

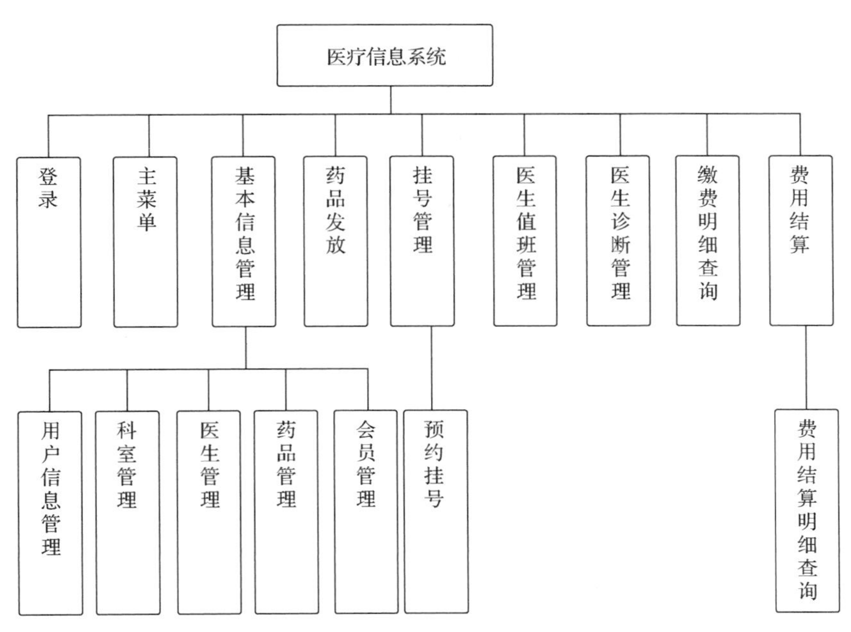

图 18-2　系统模块结构图

1. USERSSM 表

USERSSM 表用于存储使用该系统的用户信息，主键属性是 id，各个字段值的说明如表 18-1 所示。

表 18-1　USERSSM 表结构

编号	字段说明	字段名	类型	位数	小数	必需	默认值	主键	说明
1	用户 id	id	VARCHAR2	16		Y		1	
2	用户名	username	VARCHAR2	64					
3	密码	password	VARCHAR2	16					
4	角色	role	VARCHAR2	2					01 管理员 02 服务台员工 03 药剂师 04 医生 05 会员
5	真实名字	realname	VARCHAR2	64					
6	电话	tel	VARCHAR2	16					
7	年龄	age	NUMBER	10					
8	性别	sex	VARCHAR2	1					1 男，2 女
9	地址	address	VARCHAR3	128					
10	创建日期	createdate	DATE				SYSDATE		

2. MEMBERS 表

MEMBERS 表用于存储会员信息，也就是登录系统的病员信息，主键属性是 memberid，各个字段值的说明如表 18-2 所示。

表 18-2 MEMBERS 表结构

编号	字段说明	字段名	类型	位数	小数	必需	默认值	主键	说明
1	会员 id	memberid	VARCHAR2	16		Y		1	
2	会员名	name	VARCHAR2	64					
3	密码	password	VARCHAR2	16					
4	角色	role	VARCHAR2	2			05		05 会员
5	真实姓名	realname	VARCHAR2	64					
6	身份证号码	credit	VARCHAR2	64					
7	性别	sex	VARCHAR2	1					1 男，2 女
8	年龄	age	NUMBER	10					
9	电话	tel	VARCHAR2	16					
10	余额	balance	NUMBER	10	2				
11	过敏史	anaphylaxis	VARCHAR2	256					
12	创建日期	createdate	DATE				SYSDATE		

3. DEPARTMENT 表

DEPARTMENT 表用于存储科室信息，主键属性是 departid，各个字段值的说明如表 18-3 所示。

表 18-3 DEPARTMENT 表结构

编号	字段说明	字段名	类型	位数	小数	必需	默认值	主键	说明
1	科室 id	departid	VARCHAR2	16		Y		1	
2	科室名称	name	VARCHAR2	64					
3	用户 id	userid	VARCHAR3	16					创建科室的用户名
4	创建时间	createdate	DATE				SYSDATE		

4. DOCTORS 表

DOCTORS 表用于存储医生信息，主键属性是 docid，各个字段值的说明如表 18-4 所示。

表 18-4 DOCTORS 表结构

编号	字段说明	字段名	类型	位数	小数	必需	默认值	主键	说明
1	医生 id	docid	VARCHAR2	16		Y		1	
2	医生名称	name	VARCHAR2	64					
3	密码	password	VARCHAR2	16					
4	角色	role	VARCHAR3	2			04		04 医生
5	真实姓名	realname	VARCHAR4	64					
6	所属科室	departid	VARCHAR5	2					
7	职称	positional	VARCHAR6	2					
8	性别	sex	VARCHAR7	1					1 男，2 女
9	年龄	age	NUMBER						
10	电话	tel	VARCHAR9	16					
11	创建时间	createdate	DATE				SYSDATE		

5. REGISTERORDER 表

REGISTERORDER 表用于存储挂号信息，主键属性是 roid，各个字段值的说明如表 18-5 所示。

表 18-5 REGISTERORDER 表结构

编号	字段说明	字段名	类型	位数	小数	必需	默认值	主键	说明
1	挂号 id	roid	VARCHAR2	16		Y		1	
2	会员 id	memberid	VARCHAR2	16					
3	科室 id	departid	VARCHAR2	16					
4	医生 id	docid	VARCHAR2	16					
5	挂号时间	rotime	DATE				SYSDATE		
6	订单状态	status	VARCHAR2	2					01 挂号成功 02 已看病 03 已缴费 04 已发放药品
7	操作时间	optime	DATE				SYSDATE		

6. COSTSETTLEDETAILS 表

COSTSETTLEDETAILS 表用于存储费用结算明细信息，主键属性是 id，各个字段值的说明如表 18-6 所示。

表 18-6 COSTSETTLEDETAILS 表结构

编号	字段说明	字段名	类型	位数	小数	必需	默认值	主键	说明
1	id	id	VARCHAR2	16		Y		1	
2	会员 id	memberid	VARCHAR2	16					
3	挂号 id	roid	VARCHAR2	16					
4	结算金额	settleamount	NUMBER	10	2				
5	结算操作人	userid	VARCHAR2	16					
6	操作时间	createdate	DATE				SYSDATE		

7. DOCTORADVICE 表

DOCTORADVICE 表用于存储医嘱信息，主键属性是 daid，各个字段值的说明如表 18-7 所示。

表 18-7 DOCTORADVICE 表结构

编号	字段说明	字段名	类型	位数	小数	必需	默认值	主键	说明
1	医嘱 id	daid	VARCHAR2	16		Y		1	
2	挂号号	roid	VARCHAR2	16					
3	会员 id	memberid	VARCHAR2	16					
4	医生 id	docid	VARCHAR2	16					
5	诊断结果	results	VARCHAR2	128					
6	药方	prescription	VARCHAR2	256					
7	诊断时间	createdate	DATE				SYSDATE		
8	状态	status	VARCHAR2	1					删除医嘱时将该状态置为 0，正常状态为 1

8. DOCTORDUTY 表

DOCTORDUTY 表用于存储医生值班安排信息，主键属性是 ddid，各个字段值的说明如表 18-8 所示。

表 18-8　　DOCTORDUTY 表结构

编号	字段说明	字段名	类型	位数	小数	必需	默认值	主键	说明
1	值班 id	ddid	VARCHAR2	16		Y		1	
2	医生 id	docid	VARCHAR2	16					
3	值班日期	dutyday	DATE						

9. DRUGS 表

DRUGS 表用于记录卖出药品的列表信息，主键属性是 drugsid，各个字段值的说明如表 18-9 所示。

表 18-9　　DRUGS 表结构

编号	字段说明	字段名	类型	位数	小数	必需	默认值	主键	说明
1	药品 id	drugsid	VARCHAR2	16		Y		1	
2	药品名称	name	VARCHAR2	64					
3	单价	price	NUMBER	10	2				
4	进价	purchaseprice	NUMBER	10	2				
5	库存数量	num	NUMBER	15					
6	引入日期	introducedate	DATE				SYSDATE		
7	生产日期	productdate	DATE						
8	保质期	qualityperiod	DATE						
9	供货单位	supplyunit	VARCHAR2	64					
10	生产厂商	productunit	VARCHAR3	64	2				

10. DRUGSSTORRECORD 表

DRUGSSTORRECORD 表用于存储药品入库信息，主键属性是 id，各个字段值的说明如表 18-10 所示。

表 18-10　　DRUGSSTORRECORD 表结构

编号	字段说明	字段名	类型	位数	小数	必需	默认值	主键	说明
1	入库记录 id	id	VARCHAR2	16		Y		1	由序列生成
2	药品 id	drugsid	VARCHAR2	16		Y			
3	入库数量	num	NUMBER	15					
4	入库日期	storagedate	DATE				SYSDATE		

11. PARAMS 表

PARAMS 表用于存储系统中用到的参数编码信息，主键属性是 id，各个字段值的说明如表 18-11 所示。

表 18-11 PARAMS 表结构

编号	字段说明	字段名	类型	位数	小数	必需	默认值	主键	说明
1	参数 id	id	VARCHAR2	16		Y		1	
2	参数编码	p_code	VARCHAR2	64					
3	参数值	p_value	VARCHAR2	8					
4	参数名称	p_name	VARCHAR2	64					
5	父参数	parent_code	VARCHAR2	64					
6	该参数是否启用	p_flag	VARCHAR2	1					
7	显示顺序	disp_sn	VARCHAR2	8					

12. PAYMENTDETAILS 表

PAYMENTDETAILS 表用于存储缴费明细信息，主键属性是 id，各个字段值的说明如表 18-12 所示。

表 18-12 PAYMENTDETAILS 表结构

编号	字段说明	字段名	类型	位数	小数	必需	默认值	主键	说明
1	主键 id	id	VARCHAR2	16		Y		1	
2	会员 id	memberid	VARCHAR2	16					
3	充值金额	rechargeamount	NUMBER						
4	卡余额	balance	NUMBER						
5	充值方式	rechargemethod	VARCHAR2	2					
6	充值操作人员	userid	VARCHAR2	16					
7	充值发生日期	createdate	DATE				SYSDATE		

13. PRESCRIBE 表

PRESCRIBE 表用于存储开药信息，主键属性是 prescribeid，各个字段值的说明如表 18-13 所示。

表 18-13 PRESCRIBE 表结构

编号	字段说明	字段名	类型	位数	小数	必需	默认值	主键	说明
1	主键 id	prescribeid	VARCHAR2	16		Y		1	
2	医嘱 id	daid	VARCHAR2	16					
3	挂号号	roid	VARCHAR2	16					
4	药品 id	drugsid	VARCHAR2	16					
5	数量	num	NUMBER	10					
6	描述	descs	VARCHAR2	16					
7	状态	status	VARCHAR2	1					0 没发放，1 已发放

建表后数据库初始化脚本数据如下。

```
insert into MEMBERS (memberid, name, password, role, realname, credit, sex, age,
tel, balance, anaphylaxis, createdate)
values ('5', '5name', '5', '05', '会员 5', null, null, null, null, null, null,
to_date('02-01-2018', 'dd-mm-yyyy'));
insert into USERSSM (id, username, password, role, realname, tel, age, sex,
```

```
address, createdate)
values ('1', '1name', '1', '01', '管理员 1', null, null, null, null,
to_date('30-12-2017 14:27:19', 'dd-mm-yyyy hh24:mi:ss'));
insert into USERSSM (id, username, password, role, realname, tel, age, sex,
address, createdate)
values ('2', '2name', '2', '02', '服务台员工 2', null, null, null, null,
to_date('30-12-2017 14:27:19', 'dd-mm-yyyy hh24:mi:ss'));
insert into USERSSM (id, username, password, role, realname, tel, age, sex,
address, createdate)
values ('3', '3name', '3', '03', '药剂师 3', null, null, null, null,
to_date('30-12-2017 14:27:19', 'dd-mm-yyyy hh24:mi:ss'));
insert into DOCTORS (docid, name, password, role, realname, departid, positional,
sex, age, tel, address, recharge, createdate)
values ('4', '4name', '4', '04', '医生 4', null, null, null, null, null, null,
null, to_date('02-01-2018 14:25:01', 'dd-mm-yyyy hh24:mi:ss'));
insert into PARAMS (id, p_code, p_value, p_name, parent_code, p_flag, disp_sn)
values ('1', 'system_role', '01', '管理员', null, '1', null);
insert into PARAMS (id, p_code, p_value, p_name, parent_code, p_flag, disp_sn)
values ('2', 'system_role', '02', '服务台员工', null, '1', null);
insert into PARAMS (id, p_code, p_value, p_name, parent_code, p_flag, disp_sn)
values ('3', 'system_role', '03', '药剂师', null, '1', null);
insert into PARAMS (id, p_code, p_value, p_name, parent_code, p_flag, disp_sn)
values ('4', 'system_role', '04', '医生', null, '0', null);
insert into PARAMS (id, p_code, p_value, p_name, parent_code, p_flag, disp_sn)
values ('5', 'system_role', '05', '会员', null, '0', null);
insert into PARAMS (id, p_code, p_value, p_name, parent_code, p_flag, disp_sn)
values ('6', 'sex', '1', '男', null, '1', null);
insert into PARAMS (id, p_code, p_value, p_name, parent_code, p_flag, disp_sn)
values ('7', 'sex', '2', '女', null, '1', null);
insert into PARAMS (id, p_code, p_value, p_name, parent_code, p_flag, disp_sn)
values ('8', 'recharge_method', '01', '现金', null, '1', null);
insert into PARAMS (id, p_code, p_value, p_name, parent_code, p_flag, disp_sn)
values ('9', 'recharge_method', '02', '银行卡', null, '1', null);
insert into PARAMS (id, p_code, p_value, p_name, parent_code, p_flag, disp_sn)
values ('10', 'recharge_method', '03', '其他', null, '1', null);
insert into PARAMS (id, p_code, p_value, p_name, parent_code, p_flag, disp_sn)
values ('11', 'positional_titles', '01', '技师', null, '1', null);
insert into PARAMS (id, p_code, p_value, p_name, parent_code, p_flag, disp_sn)
values ('12', 'positional_titles', '02', '医师', null, '1', null);
insert into PARAMS (id, p_code, p_value, p_name, parent_code, p_flag, disp_sn)
values ('13', 'positional_titles', '03', '主任医师', null, '1', null);
insert into PARAMS (id, p_code, p_value, p_name, parent_code, p_flag, disp_sn)
values ('14', 'positional_titles', '03', '专家', null, '1', null);
```

18.2 环境搭建

18.2.1 创建工程

在 Eclipse（或 MyEclipse）中创建一个名为 HIS_SSM 的 Web Project，如图 18-3 所示。

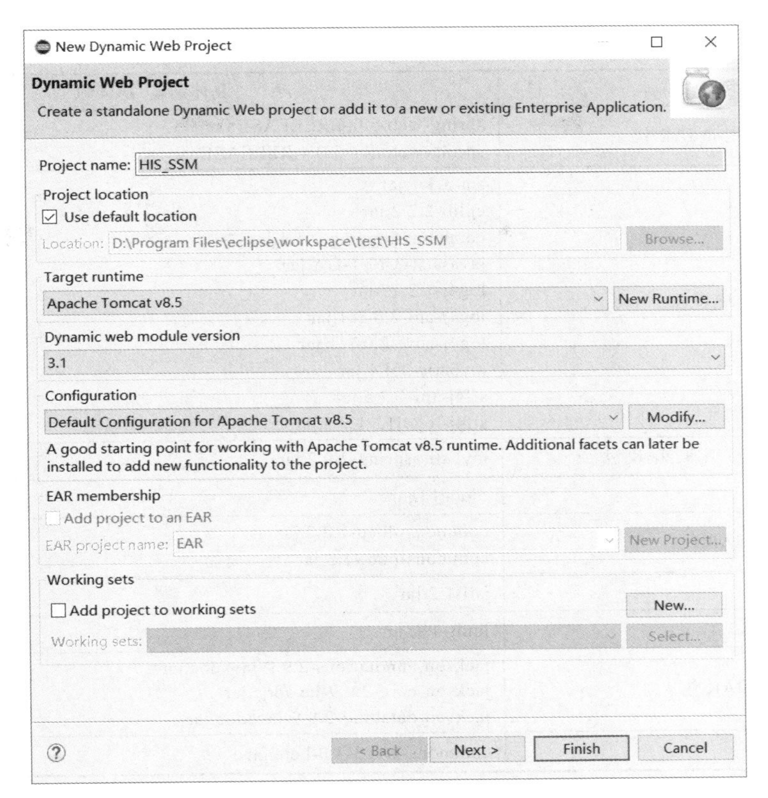

图 18-3　创建动态 Web Project

18.2.2　准备所需 JAR 包

由于本系统是使用 SSM 框架技术进行开发的，所以需要准备这三大框架的 JAR 包。除此之外，项目还涉及数据库连接、JSTL 标签库等 JAR 包。具体需要准备的 JAR 包如表 18-14 所示。

表 18-14　　　　　　　　　　　　所需 JAR 包

JAR 包分类	具体所需 JAR 包
Spring 框架所需要的 JAR 包	aopalliance-1.0.jar aspectjweaver-1.6.11.jar spring-aop-5.0.4.RELEASE.jar spring-aspects-5.0.4.RELEASE.jar spring-beans-5.0.4.RELEASE.jar spring-context-5.0.4.RELEASE.jar spring-context-indexer-5.0.4.RELEASE.jar spring-context-support-5.0.4.RELEASE.jar spring-core-5.0.4.RELEASE.jar spring-expression-5.0.4.RELEASE.jar spring-instrument-5.0.4.RELEASE.jar spring-jcl-5.0.4.RELEASE.jar spring-jdbc-5.0.4.RELEASE.jar spring-jms-5.0.4.RELEASE.jar spring-messaging-5.0.4.RELEASE.jar spring-orm-5.0.4.RELEASE.jar spring-oxm-5.0.4.RELEASE.jar spring-test-5.0.4.RELEASE.jar spring-tx-5.0.4.RELEASE.jar spring-webflux-5.0.4.RELEASE.jar spring-websocket-5.0.4.RELEASE.jar

续表

JAR 包分类	具体所需 JAR 包
Spring MVC 框架所使用的 JAR 包	spring-web-5.0.4.RELEASE.jar spring-webmvc-5.0.4.RELEASE.jar
MyBatis 框架所使用的 JAR 包	sm-3.3.1.jar cglib-2.2.2.jar commons-logging-1.1.1.jar javassist-3.17.1-GA.jar log4j-1.2.17.jar log4j-api-2.0-rc1.jar log4j-core-2.0-rc1.jar mybatis-3.4.4.jar slf4j-api-1.7.5.jar slf4j-log4j12-1.7.5.jar
MyBatis 与 Spring 整合的中间 JAR 包	mybatis-spring-2.0.2.jar
数据库驱动包	ojdbc14.jar
数据源 dbcp 所需要的 JAR 包	commons-dbcp-1.2.2.jar commons-pool-1.3.jar
JSTL 标签库 JAR 包	jstl-1.2.jar
单元测试包	junit-4.9.jar
Jackson 框架所需要的 JAR 包	jackson-annotations-2.9.9-javadoc.jar jackson-core-2.9.9-javadoc.jar jackson-databind-2.9.9-javadoc.jar
Java 工具类包	commons-lang3-3.0-beta.jar

除 commons-lang3-3.0-beta.jar 外，上面所需要的 JAR 包，都在前面使用过。读者在学习本章时，可以直接下载项目源码，并使用源码中的 JAR 包。

18.2.3 其他软件版本说明

操作系统：Windows 10。

开发工具：Eclipse Java 2018-09。

Web 服务器：Tomcat 8.5。

Java 开发包：JDK 1.8。

数据库：Oracle 12c。

浏览器：Chrome。

18.2.4 系统源码结构

为了使代码逻辑更加清晰易读，需要一个良好的系统目录结构，本医疗信息系统的目录结构如图 18-4 所示。

其中各文件包的概要说明如下。

- com.inspur.ssm.controller——存放 Controller。
- com.inspur.ssm.exception——存放异常处理文件。
- com.inspur.ssm.interceptor——存放拦截器文件。
- com.inspur.ssm.mapper——存放映射文件。
- com.inspur.ssm.pojo——存放实体类文件。
- com.inspur.ssm.service——存放服务层接口文件。

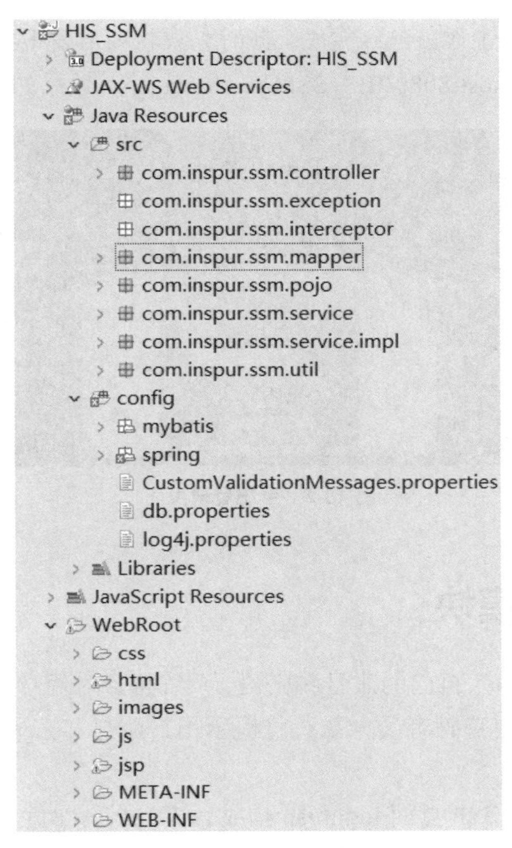

图 18-4　系统目录结构

- com.inspur.ssm.service.impl——存放服务层文件。
- com.inspur.ssm.util——存放公共工具文件。
- spring——存放 Spring 配置文件。
- mybatis——存放 MyBatis 配置文件。
- css——存放样式文件。
- image——存放图片文件。
- js——存放 JS 文件。
- jsp——存放页面文件。

18.2.5　编写配置文件

在 src 目录下创建一个源文件夹 config，并在该文件夹下分别创建数据库常量配置文件、log4j 配置文件、MyBatis 配置文件、Spring MVC 配置文件以及 Spring 配置文件。所有的配置文件与第 17 章讲解 SSM 整合时的配置代码基本相同（注意修改数据库信息和包名称），这里不再重复。

18.2.6　引入页面资源

将项目运行所需要的字体、图片、CSS 格式、JS 和 JSP 文件按照图 18-4 所示的结构引入项目中。页面所需要引入的文件可以从下载的源码文件中获得。

至此，系统开发的前期工作就已经完成。此时如果将项目发布到 Tomcat 8.5 服务器上并访问项目首页地址 http://localhost:8080/HIS_SSM/jsp/login.jsp，效果如图 18-5 所示。

图 18-5　系统登录页面

18.3　用户登录模块

用户可以在该模块中输入自己的账号和密码，系统将会对用户账号和用户密码进行验证，如果输入的用户账号和用户密码有误，系统将提示用户账号和密码不正确，无法进入系统主界面。

该模块视图部分由一个 JSP 页面 login.jsp 构成，该 JSP 页面负责提交用户的登录信息到控制器，并进入系统主界面。POJO 对象 Userssm 负责存储用户信息。控制器 LoginController 负责分发用户请求至后台。Service 服务层 LoginService 负责业务逻辑处理。DAO 层 UserssmMapper 负责连接数据库，获取数据信息。

18.3.1　视图

视图部分由一个 JSP 页面 login.jsp 构成。login.jsp 页面负责提供输入登录信息界面，并负责显示登录反馈信息，效果参见图 18-5。

程序清单：HIS_SSM/WebRoot/jsp/login.jsp。

```
<%@ page language="java" contentType="text/html; charset=UTF-8"
pageEncoding="UTF-8"%>
<%String path = request.getContextPath();%>
<!DOCTYPE HTML>
<html>
<head>
<meta http-equiv="Content-Type" content="text/html; charset=utf-8" />
<title>社区医疗管理系统</title>
<link href="<%=path %>/css/login.css" rel="stylesheet" type="text/css" />
<script type="text/javascript" src="<%=path %>/js/jquery.min.js"></script>
<script type="text/javascript">
    $(function(){
        $('.captcha').focus(function(){
            $('.yzm-box').show();
        });
```

```
        $('.captcha').focusout(function(){
            $('.yzm-box').hide();
        });
    });
</script>
</head>
<body>
<div id="message-box"> 用户名或密码错误！ </div>
<div id="wrap">
    <div id="header"></div>
<div id="content-wrap">
    <div class="space"></div>
    <form action="<%=path %>/login.action" method="post" name="ThisForm">
        <div class="content">

    <div class="field"><label>账　户: </label><input class="username"
name="username"  type="text" /></div>
              <div class="field"><label>密　码:</label><input class="password"
name="password"  type="password" /><br /></div>
    <div class="field"><span style="color: red">${message}</span></div>
    </div>
<div class="btn"><input type="submit" class="login-btn" value="" /></div>
</form>
</div>
<div id="footer"></div>
</div>
</body>
</html>
```

18.3.2　POJO 类

POJO 类用于描述用户的重要信息，在该类中可定义用户相关属性以及相应的 getter、setter 方法。

程序清单：HIS_SSM/src/com/inspur/ssm/pojo/Userssm.java。

```
package com.inspur.ssm.pojo;

import java.math.BigDecimal;
import java.util.Date;

public class Userssm {
    private String id;
    private String username;
    private String password;
    private String role;
    private String realname;
    private String tel;
    private BigDecimal age;
    private String sex;
    private String address;
    private Date createdate;
    public String getId() {
        return id;
```

```
    }
    public void setId(String id) {
         this.id = id == null ? null : id.trim();
    }
    public String getUsername() {
         return username;
    }
    public void setUsername(String username) {
         this.username = username == null ? null : username.trim();
    }
    public String getPassword() {
         return password;
    }
    public void setPassword(String password) {
         this.password = password == null ? null : password.trim();
    }
    public String getRole() {
         return role;
    }
    public void setRole(String role) {
         this.role = role == null ? null : role.trim();
    }
    public String getRealname() {
         return realname;
    }
    public void setRealname(String realname) {
         this.realname = realname == null ? null : realname.trim();
    }
    public String getTel() {
         return tel;
    }
    public void setTel(String tel) {
         this.tel = tel == null ? null : tel.trim();
    }
    public BigDecimal getAge() {
         return age;
    }
    public void setAge(BigDecimal age) {
         this.age = age;
    }
    public String getSex() {
         return sex;
    }
    public void setSex(String sex) {
         this.sex = sex == null ? null : sex.trim();
    }
    public String getAddress() {
         return address;
    }
    public void setAddress(String address) {
         this.address = address == null ? null : address.trim();
    }
    public Date getCreatedate() {
         return createdate;
    }
```

```
    public void setCreatedate(Date createdate) {
        this.createdate = createdate;
    }
}
```

18.3.3 控制器

控制器 LoginController 负责分发客户请求，将请求分发到对应的 Service 进行处理，并对处理结果进行判断处理。该 Controller 类中主要包含 3 个处理方法，分别是登录、退出和修改密码。

程序清单：HIS_SSM/src/com/inspur/ssm/controller/LoginController.java。

```
package com.inspur.ssm.controller;

import javax.servlet.http.HttpServletRequest;
import javax.servlet.http.HttpSession;

import org.springframework.beans.factory.annotation.Autowired;
import org.springframework.stereotype.Controller;
import org.springframework.web.bind.annotation.RequestMapping;

import com.inspur.ssm.pojo.Userssm;
import com.inspur.ssm.service.LoginService;
/**
 * 登录控制器
 * @author inspur
 *
 */
@Controller
public class LoginController {
    @Autowired
    private LoginService loginService;
    //登录
    @RequestMapping("/login")
    public String login(HttpServletRequest request, Userssm userssm) throws
        Exception{
        Userssm user = null;
        try{
            //调用 service 进行用户身份验证
            user = loginService.findUser(userssm);
        }catch(Exception e){
            e.printStackTrace();
            request.setAttribute("message", e.getMessage());
        }
        if(user!=null){
            HttpSession session = request.getSession();
            //在 session 中保存用户身份信息
            session.setAttribute("user", user);
            //转发到商品列表中
            return "index";
        }

        return "login";
    }
```

```
        //退出
        @RequestMapping("/logout")
        public String logout(HttpSession session)throws Exception {
            // 清除 session，设置 session 过期
            session.invalidate();
            return "login";
        }
        //修改密码
        @RequestMapping("/updatePwd")
        public String updatePwd(HttpServletRequest request,HttpSession
            session)throws Exception {
            //原密码
            String password1 = request.getParameter("password1");
            //新密码
            String password2 = request.getParameter("password2");
            //新密码确认
            String password3 = request.getParameter("password3");
            //如果有任何一个值为空，则返回错误信息
            if(password1 == null || "".equals(password1)
                || password2 == null || "".equals(password2)
                || password3 == null || "".equals(password3)){
            request.setAttribute("message", "请输入原始密码、新密码和确认密码信息!");
                return "userpwdupdate";
            }
            //新密码和新密码确认不相同，返回错误信息
            if(!password2.equals(password3)){
                request.setAttribute("message", "请输入相同的新密码和确认密码!");
                return "userpwdupdate";
            }
            //取出当前登录对象
            Userssm user = (Userssm) session.getAttribute("user");
            if(user != null){
                if(!password1.equals(user.getPassword()) ){
                    request.setAttribute("message", "请输入正确的原密码!");
                    return "userpwdupdate";
                }
            }
            //更新密码
            user.setPassword(password2);
            loginService.updateByPrimaryKey(user);
            // 清除 session，设置 session 过期
            session.invalidate();
            return "login";
        }
    }
```

18.3.4 服务层

服务层通过接口的方式实现多态，减少了模块之间的耦合，使各个成员依赖于抽象，而不是依赖于具体，方便了程序的维护和扩展。服务层的接口是 LoginService，LoginServiceImpl 是具体的实现类。服务层主要负责业务逻辑处理，如用户登录时判断用户是否存在。其主要包含

两个处理方法：查询用户信息和更新用户信息。

程序清单：HIS_SSM/src/com/inspur/ssm/service/LoginService.java。

```
package com.inspur.ssm.service;

import javax.annotation.Resource;

import com.inspur.ssm.pojo.Userssm;

/**
 * 登录服务类
 * @author inspur
 *
 */
@Resource
public interface LoginService {
    //用户查询列表
    public Userssm findUser(Userssm userssm)throws Exception;

    public void updateByPrimaryKey(Userssm user);
}
```

程序清单：HIS_SSM/src/com/inspur/ssm/service/impl/LoginServiceImpl.java。

```
package com.inspur.ssm.service.impl;

import org.springframework.beans.factory.annotation.Autowired;
import org.springframework.stereotype.Service;

import com.inspur.ssm.mapper.UserssmMapper;
import com.inspur.ssm.pojo.Userssm;
import com.inspur.ssm.service.LoginService;
@Service("loginService")
public class LoginServiceImpl implements LoginService{

    @Autowired
    private UserssmMapper userssmMapper;

    @Override
    public Userssm findUser(Userssm userssmQuery) throws Exception {
        Userssm userssm = userssmMapper.findUser(userssmQuery);
        if(userssm == null){
            throw new Exception("用户信息不存在");
        }else{
            return userssm;
        }
    }

    @Override
    public void updateByPrimaryKey(Userssm user) {
        userssmMapper.updateByPrimaryKey(user);
    }
}
```

18.3.5 持久层

持久层负责连接数据库、获取数据信息、将用户提交的信息写入数据库等功能。该层通过接口和映射文件的方式实现，接口是 UserssmMapper.java，映射文件为 UserssmMapper.xml。

程序清单：HIS_SSM/src/com/inspur/ssm/mapper/UserssmMapper.java。

```
package com.inspur.ssm.mapper;

import java.util.List;
import java.util.Map;

import com.inspur.ssm.pojo.Userssm;

public interface UserssmMapper {
     int deleteByPrimaryKey(String id);
     int insert(Userssm record);
     int insertSelective(Userssm record);
     Userssm selectByPrimaryKey(String id);
     int updateByPrimaryKeySelective(Userssm record);
     int updateByPrimaryKey(Userssm record);
     Userssm findUser(Userssm userssmQuery);
     public List<Userssm> getUserList(Map<String,Object> params);
     //分页总条数
     public Long getCounts(Map<String,Object> p);
}
```

程序清单：HIS_SSM/src/com/inspur/ssm/mapper/UserssmMapper.xml。

```
<?xml version="1.0" encoding="UTF-8" ?>
<!DOCTYPE mapper PUBLIC "-//mybatis.org//DTD Mapper 3.0//EN"
"http://mybatis.org/dtd/mybatis-3-mapper.dtd" >
<mapper namespace="com.inspur.ssm.mapper.UserssmMapper" >
<resultMap id="BaseResultMap" type="com.inspur.ssm.pojo.Userssm" >
<id column="ID" property="id" jdbcType="VARCHAR" />
<result column="USERNAME" property="username" jdbcType="VARCHAR" />
<result column="PASSWORD" property="password" jdbcType="VARCHAR" />
<result column="ROLE" property="role" jdbcType="VARCHAR" />
<result column="REALNAME" property="realname" jdbcType="VARCHAR" />
<result column="TEL" property="tel" jdbcType="VARCHAR" />
<result column="AGE" property="age" jdbcType="DECIMAL" />
<result column="SEX" property="sex" jdbcType="VARCHAR" />
<result column="ADDRESS" property="address" jdbcType="VARCHAR" />
<result column="CREATEDATE" property="createdate" jdbcType="DATE" />
</resultMap>
<sql id="Base_Column_List" >
     ID, USERNAME, PASSWORD, ROLE, REALNAME, TEL, AGE, SEX, ADDRESS, CREATEDATE
</sql>

<sql id="query_user_where">
     <if test="id != '' and id != null">
          and t.id like '%${id}%'
     </if>
     <if test="username != ''  and username != null">
          and t.username like '%${username}%'
```

```
        </if>
        <if test="role != '' and role != null">
            and t.role = '${role}'
        </if>
    </sql>


    <select id="getCounts" resultType="long" parameterType="java.util.HashMap">
        select count(*) from USERSSM  t
        <where>
            <include refid="query_user_where"></include>
        </where>
    </select>

    <select id="getUserList"  resultType="java.util.List" resultMap=" BaseResultMap"
        parameterType="java.util.HashMap">
    <include refid="com.inspur.ssm.mapper.CommonMapper.pagingTop"></include>
        select * from USERSSM    t
        <where>
           <include refid="query_user_where"></include>
        </where>
    <include refid="com.inspur.ssm.mapper.CommonMapper.pagingBottom"></include>
    </select>


    <select id="findUser" resultType="com.inspur.ssm.pojo.Userssm"
    parameterType="com.inspur.ssm.pojo.Userssm" >
         select t1.id,t1.username,t1.role,t1.realname,t1.password from USERSSM t1
            where username = #{username} and password = #{password}
         union
         select t2.docid,t2.name,t2.role,t2.realname,t2.password from doctors t2
            where name = #{username} and password = #{password}
         union
         select t3.memberid,t3.name,t3.role,t3.realname,t3.password from members t3
            where name = #{username} and password = #{password}
    </select>

    <select id="selectByPrimaryKey" resultMap="BaseResultMap"
         parameterType="java.lang.String" >
         select
    <include refid="Base_Column_List" />
         from USERSSM
         where ID = #{id,jdbcType=VARCHAR}
    </select>
    <delete id="deleteByPrimaryKey" parameterType="java.lang.String" >
         delete from USERSSM
         where ID = #{id,jdbcType=VARCHAR}
    </delete>
    <insert id="insert" parameterType="com.inspur.ssm.pojo.Userssm" >
        insert into USERSSM (ID, USERNAME, PASSWORD,
        ROLE, REALNAME, TEL,
        AGE, SEX, ADDRESS,
        CREATEDATE)
        values (#{id,jdbcType=VARCHAR}, #{username,jdbcType=VARCHAR},
```

```
        #{password,jdbcType=VARCHAR},
        #{role,jdbcType=VARCHAR}, #{realname,jdbcType=VARCHAR}, #{tel,jdbcType=
        VARCHAR},
        #{age,jdbcType=DECIMAL}, #{sex,jdbcType=VARCHAR}, #{address,jdbcType=
        VARCHAR},
        #{createdate,jdbcType=DATE})
    </insert>
    <insert id="insertSelective" parameterType="com.inspur.ssm.pojo.Userssm" >
        insert into USERSSM
    <trim prefix="(" suffix=")" suffixOverrides="," >
    <if test="id != null" >
         ID,
    </if>
    <if test="username != null" >
         USERNAME,
    </if>
    <if test="password != null" >
         PASSWORD,
    </if>
    <if test="role != null" >
         ROLE,
    </if>
    <if test="realname != null" >
         REALNAME,
    </if>
    <if test="tel != null" >
         TEL,
    </if>
    <if test="age != null" >
         AGE,
    </if>
    <if test="sex != null" >
         SEX,
    </if>
    <if test="address != null" >
        ADDRESS,
    </if>
    <if test="createdate != null" >
        CREATEDATE,
    </if>
    </trim>
    <trim prefix="values (" suffix=")" suffixOverrides="," >
    <if test="id != null" >
        #{id,jdbcType=VARCHAR},
    </if>
    <if test="username != null" >
        #{username,jdbcType=VARCHAR},
    </if>
    <if test="password != null" >
        #{password,jdbcType=VARCHAR},
    </if>
    <if test="role != null" >
        #{role,jdbcType=VARCHAR},
    </if>
```

```
    <if test="realname != null" >
        #{realname,jdbcType=VARCHAR},
    </if>
    <if test="tel != null" >
        #{tel,jdbcType=VARCHAR},
    </if>
    <if test="age != null" >
        #{age,jdbcType=DECIMAL},
    </if>
    <if test="sex != null" >
        #{sex,jdbcType=VARCHAR},
    </if>
    <if test="address != null" >
        #{address,jdbcType=VARCHAR},
    </if>
    <if test="createdate != null" >
        #{createdate,jdbcType=DATE},
    </if>
    </trim>
    </insert>
    <update id="updateByPrimaryKeySelective"
    parameterType="com.inspur. ssm.pojo.Userssm" >
        update USERSSM
    <set >
    <if test="username != null" >
        USERNAME = #{username,jdbcType=VARCHAR},
    </if>
    <if test="password != null" >
        PASSWORD = #{password,jdbcType=VARCHAR},
    </if>
    <if test="role != null" >
        ROLE = #{role,jdbcType=VARCHAR},
    </if>
    <if test="realname != null" >
        REALNAME = #{realname,jdbcType=VARCHAR},
    </if>
    <if test="tel != null" >
        TEL = #{tel,jdbcType=VARCHAR},
    </if>
    <if test="age != null" >
        AGE = #{age,jdbcType=DECIMAL},
    </if>
    <if test="sex != null" >
        SEX = #{sex,jdbcType=VARCHAR},
    </if>
    <if test="address != null" >
        ADDRESS = #{address,jdbcType=VARCHAR},
    </if>
    <if test="createdate != null" >
        CREATEDATE = #{createdate,jdbcType=DATE},
    </if>
    </set>
        where ID = #{id,jdbcType=VARCHAR}
    </update>
```

```
    <update id="updateByPrimaryKey" parameterType="com.inspur.ssm.pojo.Userssm" >
        update USERSSM
        set USERNAME = #{username,jdbcType=VARCHAR},
        PASSWORD = #{password,jdbcType=VARCHAR},
        ROLE = #{role,jdbcType=VARCHAR},
        REALNAME = #{realname,jdbcType=VARCHAR},
        TEL = #{tel,jdbcType=VARCHAR},
        AGE = #{age,jdbcType=DECIMAL},
        SEX = #{sex,jdbcType=VARCHAR},
        ADDRESS = #{address,jdbcType=VARCHAR},
        CREATEDATE = #{createdate,jdbcType=DATE}
        where ID = #{id,jdbcType=VARCHAR}
    </update>
    </mapper>
```

18.3.6 启动项目测试登录

将项目发布到 Tomcat 服务器并启动，成功访问登录页面后，即可输入用户账号和密码登录系统。登录成功后主界面如图 18-6 所示。单击系统主界面右上角的“注销”图标即可退出登录，单击“修改密码”图标即可进行用户密码的修改操作。

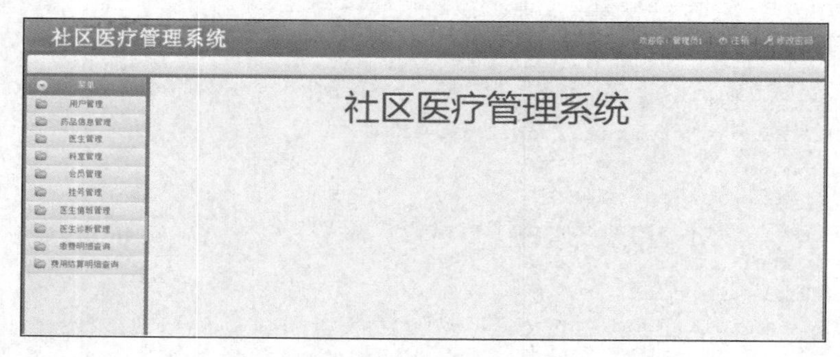

图 18-6 系统主界面

18.4 用户管理模块

用户管理模块是本系统的核心模块之一，该模块实现了对使用系统用户的查询、添加、修改和删除功能。下面以查询用户信息为例进行展示。页面显示如图 18-7 所示。

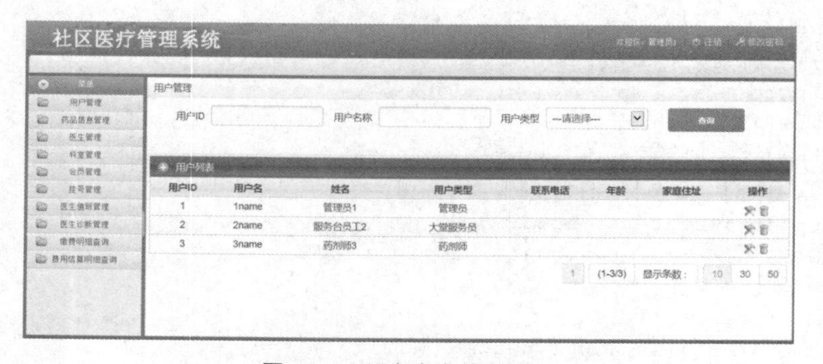

图 18-7 用户查询信息页面

18.4.1　页面显示

程序清单：HIS_SSM/WebRoot/jsp/adminList.jsp。

```
<%@ page language="java" contentType="text/html; charset=UTF-8"
    pageEncoding="UTF-8"%>
<%@ include file="/jsp/common/includehead.jsp"%>
<!DOCTYPE html>
<html xmlns="http://www.w3.org/1999/xhtml">
<head>
<meta http-equiv="Content-Type" content="text/html; charset=utf-8" />
<link href="${ctx}/css/select.css" rel="stylesheet" type="text/css" />
<link href="${ctx}/css/bootstrap.min.css" rel="stylesheet" type="text/css">
<script type="text/javascript" src="${ctx}/js/select-ui.min.js"></script>
<script type="text/javascript">
    function deletes(id){
        window.location.href="adminlist.html";
    }
    function modify(id){
        window.location.href="adminadd.html";
    }
    function query(){
        window.location.href="queryUserList.action";
    }
</script>
</head>

<body>
<div id="contentWrap">
    <!--表格控件 -->
<div id="widget table-widget">
<div class="pageTitle">用户管理</div>
<div class="querybody" >
<form name="queryForm" action="${ctx}/user/queryUserList.action"
    method="post">
    <ul class="seachform" >
    <li><label>用户 ID</label><input id="id" name="id" type="text"
        value="${userssm.id}" class="scinput" /></li>
    <li><label>用户名称</label><input id="username" name="username" type="text"
        value="${userssm.username}" class="scinput" /></li>
    <li><label>用户类型</label>
        <select id="role" name="role" style="width:150px;height:32px;" >
            <c:forEach items="${roleList}" var="role" >
                <option value="${role.BM}" <c:if test="${userssm.role
                    eqrole.BM}">selected</c:if>>${role.MC}</option>
            </c:forEach>
        </select>
    </li>
    <li><label> </label><input type="submit" class="scbtn" value="查询" />
    </li>
    </ul>
    </form>
    </div>
<div class="pageColumn">
```

```
<div class="pageButton"><a href="${ctx}/html/adminadd.html"><img src="${ctx}/
images/t01.png" title="新增"/></a><span>用户列表</span></div>
<table>
<thead>
     <th width="">用户 ID</th>
     <th width="">用户名</th>
     <th width="">姓名</th>
     <th width="">用户类型</th>
     <th width="">联系电话</th>
     <th width="">年龄</th>
     <th width="">家庭住址</th>
     <th width="10%">操作</th>
</thead>
<tbody>
<c:forEach items="${userList}" var="user">
     <tr>
          <td>${user.id}</td>
          <td>${user.username}</td>
          <td>${user.realname}</td>
          <td>${user.role eq 01 ? "管理员":user.role eq 02 ? "大堂服务员":
         user.role eq 03 ? "药剂师": "未知"}</td>
          <td>${user.tel}</td>
          <td>${user.age}</td>
          <td>${user.address}</td>
          <td><a onclick="modify(id)"><img src="${ctx}/images/icon/
               edit.png" width="16" height="16" /></a>
           <a onclick="deletes(id)"><img src="${ctx}/images/icon/del.png"
               width="16" height="16" /></a></td>
           </tr>
           </c:forEach>
</tbody>
</table>
<jsp:include page="common/includefoot.jsp">
    <jsp:param name="url" value="/user/queryUserList.action" />
    </jsp:include>
</div>
</div><!-- #widget -->
</div>
</body>
</html>
```

18.4.2 POJO 类

此处程序与登录模块的一样。

程序清单：HIS_SSM/src/com/inspur/ssm/pojo/Userssm.java。

18.4.3 控制器

程序清单：HIS_SSM/src/com/inspur/ssm/controller/UserController.java。

```
package com.inspur.ssm.controller;
```

```
import java.util.HashMap;
import java.util.List;
import java.util.Map;

import javax.servlet.http.HttpServletRequest;

import org.springframework.beans.factory.annotation.Autowired;
import org.springframework.stereotype.Controller;
import org.springframework.web.bind.annotation.ModelAttribute;
import org.springframework.web.bind.annotation.RequestMapping;

import com.inspur.ssm.pojo.Userssm;
import com.inspur.ssm.service.UserService;
import com.inspur.ssm.util.CommonUtil;
import com.inspur.ssm.util.Page;
import com.inspur.ssm.util.StringUtil;

@Controller
@RequestMapping("/user")
public class UserController extends PageController{

    @Autowired
    private UserService userService;

    @RequestMapping("/queryUserList")
    public String findUserList(HttpServletRequest
        request,@ModelAttribute("userssm") Userssm userssm){
        List<Map<String,Object>> roleList = CommonUtil.getCode("system_role",
            true, "asc");
        Map<String,Object> params = new HashMap<String,Object>();
        //添加查询条件
        if(!StringUtil.isEmptyString(userssm.getId())){
            params.put("id",userssm.getId());
        }
        if(!StringUtil.isEmptyString(userssm.getUsername())){
            params.put("username",userssm.getUsername());
        }
        if(!StringUtil.isEmptyString(userssm.getRole())){
            params.put("role",userssm.getRole());
        }

        //获取总条数
        Long totalCount = userService.getCounts(params);
        //设置分页对象
        Page page = executePage(request,totalCount);
        //如排序
        if(page.isSort()){
            params.put("orderName",page.getSortName());
            params.put("descAsc",page.getSortState());
        }else{
            //没有进行排序，默认排序方式
            params.put("orderName","id");
            params.put("descAsc","desc");
        }
```

```
            //压入查询参数：开始条数与结束条灵敏
            params.put("startIndex", page.getBeginIndex());
            params.put("endIndex", page.getEndinIndex());

            //查询集合
            List<Userssm> users = userService.getUserList(params);
            request.setAttribute("userList",users);
            request.setAttribute("roleList", roleList);
            return "adminList";
        }
    }
```

18.4.4 服务层

此外代码同样分为接口 UserService 和实现类 UserServiceImpl。

程序清单：HIS_SSM/src/com/inspur/ssm/service/UserService.java。

```
package com.inspur.ssm.service;

import java.util.List;
import java.util.Map;

import javax.annotation.Resource;
import com.inspur.ssm.pojo.Userssm;
@Resource
public interface UserService {
    public Long getCounts(Map<String, Object> params);
    public List<Userssm> getUserList(Map<String, Object> params);

}
```

程序清单：HIS_SSM/src/com/inspur/ssm/service/impl/UserServiceImpl.java。

```
package com.inspur.ssm.service.impl;

import java.util.List;
import java.util.Map;

import org.springframework.beans.factory.annotation.Autowired;
import org.springframework.stereotype.Service;

import com.inspur.ssm.mapper.UserssmMapper;
import com.inspur.ssm.pojo.Userssm;
import com.inspur.ssm.service.UserService;
@Service("userService")
public class UserServiceImpl implements UserService{
    @Autowired
    private UserssmMapper userssmMapper;

    @Override
    public Long getCounts(Map<String, Object> params) {
        return userssmMapper.getCounts(params);
    }

    @Override
```

```
        public List<Userssm> getUserList(Map<String, Object> params) {
            return userssmMapper.getUserList(params);
        }
    }
```

18.4.5　持久层

此处代码与登录模块一致，分为接口 UserssmMapper.java 和映射文件 UserssmMapper.xml。

程序清单：HIS_SSM/src/com/inspur/ssm/mapper/UserssmMapper.java；HIS_SSM/src/com/inspur/ssm/mapper/impl/ UserssmMapper.xml。

本章小结

本章使用 SSM 框架开发了一个 Java EE 项目——医疗信息管理系统。因为企业平台本身的复杂性，所以本项目涉及的模块较多，而且业务逻辑也比较复杂，这些对初学者来说可能会有一定难度，但只要读者先认真阅读本书前 17 章所介绍的知识，并结合本章的讲解，就一定可以掌握本章介绍的内容。

本章介绍的 Java EE 应用综合了前面 17 章所介绍的知识，因此本章内容既是对前面知识的回顾和复习，也是将理论知识应用到实际开发的典范。一旦读者掌握了本章案例的开发方法后，就会对实际 Java EE 企业应用的开发产生豁然开朗的感觉。